Sustainable Textiles: Production, Processing, Manufacturing & Chemistry

Series Editor

Subramanian Senthilkannan Muthu, Chief Sustainability Officer, Green Story Inc., Tung Chung, Hong Kong

This series aims to address all issues related to sustainability through the lifecycles of textiles from manufacturing to consumer behavior through sustainable disposal. Potential topics include but are not limited to: Environmental Footprints of Textile manufacturing; Environmental Life Cycle Assessment of Textile production; Environmental impact models of Textiles and Clothing Supply Chain; Clothing Supply Chain Sustainability; Carbon, energy and water footprints of textile products and in the clothing manufacturing chain; Functional life and reusability of textile products; Biodegradable textile products and the assessment of biodegradability; Waste management in textile industry; Pollution abatement in textile sector; Recycled textile materials and the evaluation of recycling; Consumer behavior in Sustainable Textiles; Eco-design in Clothing & Apparels; Sustainable polymers & fibers in Textiles; Sustainable waste water treatments in Textile manufacturing; Sustainable Textile Chemicals in Textile manufacturing. Innovative fibres, processes, methods and technologies for Sustainable textiles; Development of sustainable, eco-friendly textile products and processes; Environmental standards for textile industry; Modelling of environmental impacts of textile products; Green Chemistry, clean technology and their applications to textiles and clothing sector; Eco-production of Apparels, Energy and Water Efficient textiles. Sustainable Smart textiles & polymers, Sustainable Nano fibers and Textiles; Sustainable Innovations in Textile Chemistry & Manufacturing; Circular Economy, Advances in Sustainable Textiles Manufacturing; Sustainable Luxury & Craftsmanship; Zero Waste Textiles.

Canan Saricam • Nazan Okur

Changing Textile and Apparel Consumption in Transformative Era of Sustainability and Digitalization

Springer

Canan Saricam
Department of Textile Engineering
Faculty of Textile Technologies and Design,
Istanbul Technical University
Istanbul, Türkiye

Nazan Okur
Department of Textile Engineering
Faculty of Textile Technologies and Design,
Istanbul Technical University
Istanbul, Türkiye

ISSN 2662-7108 ISSN 2662-7116 (electronic)
Sustainable Textiles: Production, Processing, Manufacturing & Chemistry
ISBN 978-3-031-97418-2 ISBN 978-3-031-97419-9 (eBook)
https://doi.org/10.1007/978-3-031-97419-9

This Springer imprint is published by the registered company Springer Nature Switzerland AG
The registered company address is: Gewerbestrasse 11, 6330 Cham, Switzerland

Contents

Introduction

Keywords Sustainability · Digital transformation · Digitalization · Textile · Apparel · Fashion retailing · Consumer · Fashion consumption · Modern tools · E-commerce · Mobile shopping · Social marketing · COVID-19

The concept of sustainability emerged in the second half of the twentieth century and has evolved over time with the contributions of stakeholders such as governments, academia, companies, brands, and consumers. Today's critical problems such as global warming, environmental pollution, and depletion of natural resources are directing countries and institutions towards a more sustainable development model. In 1987, the Brundtland Report determined the four basic principles of sustainability, and in 1997, John Elkington's Triple Bottom Line approach defined the environmental, social and economic dimensions (Correia, 2019). In 2015, the United Nations (UN) determined the Sustainable Development Goals (SDGs) and ensured their universal adoption. UN aims to provide global sustainability in social, economic and environmental areas with the 17 SDGs. Issues such as combating climate change, sustainable production and consumption, transition to clean energy, and sustainable transformation of industry and infrastructure guide both governments and the private sector in policy-making processes. Sustainability practices in various sectors such as textile and apparel, automotive, construction and energy are developing in line with the goals set by the UN (https://sdgs.un.org/goals).

The textile and apparel industries are among the most resource-intensive sectors in the world; requiring vast amounts of freshwater, being responsible for significant levels of carbon emissions, and generating large quantities of waste along their supply chains. As consumers become more conscious of environmental and ethical issues and laws and regulations become stricter, both manufacturers and brands are responding by introducing more sustainable materials, production methods, and supply chain processes to reduce their environmental impact and meet the demand for ethically produced, eco-friendly products. In parallel with this, European Union (EU) strategy for sustainable and circular textiles addresses textile production and

C. Saricam, N. Okur, *Changing Textile and Apparel Consumption in Transformative Era of Sustainability and Digitalization*, Sustainable Textiles: Production, Processing, Manufacturing & Chemistry,
https://doi.org/10.1007/978-3-031-97419-9_1

consumption by implementing the commitments of the European Green Deal, the Circular Economy Action Plan, and the European Industrial Strategy to create a greener and more competitive textile industry. The EU is leading the spread of environmentally friendly policies at a global level and promoting sustainable production and consumption models. The EU Strategy for Sustainable and Circular Textiles presents a plan of action for the future, including the creation of design features that increase the lifespan of textiles, ease of repair and recycling and minimum amount of recycled material, the implementation of a Digital Product Passport (DPP), preventing disposal of products left in stock or returned, tackling the release of microplastics from manmade fibers, combating greenwashing, limiting the export of textile waste (Waste Shipment Regulation, proposed in 2021), introducing Extended Producer Responsibility (EPR) rules for textiles and encouraging manufacturers to design more sustainable products, promoting circular economy models (rental, second-hand, etc.) and ensuring that consumers benefit from high-quality affordable textiles for longer (https://environment.ec.europa.eu/strategy/textiles-strategy_en#related-strategies).

The European Green Deal, introduced in 2019, aims for the continent to be carbon neutral by 2050. In this context, several targets have been set, such as reducing greenhouse gas emissions, investing in renewable energy and expanding circular economy practices. Under the Green Deal, the Carbon Border Adjustment Mechanism (CBAM) encourages low-carbon production by introducing additional costs for importing carbon-intensive products, while the Circular Economy Action Plan announces initiatives across the entire life cycle of products (https://commission.europa.eu/strategy-and-policy/priorities-2019-2024/european-green-deal_en). In addition to promoting sustainable consumption and circular economy practices, the Circular Economy Action Plan focuses on product design and aims to prevent waste and extend resource use in the EU economy by ensuring that products have a longer lifespan, are repairable, and are recyclable (https://environment.ec.europa.eu/strategy/circular-economy-action-plan_en). As an implementation for this purpose, Ecodesign for Sustainable Products Regulation (ESPR), proposed in 2022, aims to make products offered to the EU market more environmentally friendly, energy-efficient, and recyclable. Key regulations introduced by the ESPR include increasing the durability, repairability and recyclability of products, tracking the raw material, manufacturing process and recycling potential of each product through digital product passports, and requiring manufacturers to make more transparent environmental declarations to prevent greenwashing. The ESPR will directly affect industries such as textiles and apparel, electronics, furniture, and construction materials, making sustainable production mandatory in the European market (https://commission.europa.eu/energy-climate-change-environment/standards-tools-and-labels/products-labelling-rules-and-requirements/ecodesign-sustainable-products-regulation_en).

Likewise, in the United States, there are important strategies and regulations focusing on issues such as combating climate change, clean energy transition, and circular economy. The United States has rejoined the Paris Climate Agreement and committed to net-zero emissions by 2050. To that end, several green initiatives are being implemented at the federal and state levels (https://unfccc.int/sites/default/files/resource/US_accessibleLTS2021.pdf). The National Recycling Strategy, launched by the US Environmental Protection Agency (EPA), aims to increase recycling rates and reduce plastic waste by 2030 (https://www.epa.gov/circulareconomy). Fashion Sustainability and Social Accountability Act in New York and Responsible Textile Recovery Act in California aim to reduce textile waste and increase recycling practices (https://www.fashiondive.com/news/US-state-legislatures-fashion-sustainability-safety/738813/). These initiatives and regulations are accelerating the sustainable transformation in the textile, apparel, and fashion industries. In addition to the US and the EU, countries such as China, India, Turkey, Bangladesh, Brazil, and Japan have significant sustainability-related regulations and programs in the textile and apparel sector. For example, in the countries and regions mentioned above, Extended Producer Responsibility (EPR) is implemented as a policy framework that holds producers responsible for their products throughout their entire life cycle, including after consumption (https://www.oecd.org/en/topics/extended-producer-responsibility-and-economic-instruments.html). These countries also offer various obligations and incentives on issues such as carbon reduction, waste management, water conservation, chemical use restrictions, and recycling, and implement circular economy and zero-waste programs.

Simultaneously, advances in technology have brought innovations that have fundamentally transformed the fashion industry. Eventually, sustainability and digital transformation are interrelated and complementary concepts in the textile and apparel industries. Both are essential components that are in the process of shaping the future of industries, and, in addition, digital transformation is utilized as an essential mechanism to advance sustainability goals, offering many solutions to support sustainability. Sustainability and digital transformation together provide the industry to be more sustainable, transparent, efficient, flexible, innovative, and consumer-focused. As part of the UN Sustainable Development Goals, the fashion industry is trying to become more sustainable by adopting digital transformation that aligns with Goal 12 (Responsible Consumption and Production) and Goal 13 (Climate Action). For example, fashion designers and brands are reducing costs and their environmental footprint, with 3D design and virtual fashion shows. Artificial intelligence and big data analytics are also playing a critical role in predicting fashion trends and improving inventory management. Within the framework of Sustainable Products Initiative in 2022, the EU has started working towards transparency and digitizing the life cycle of products (https://ec.europa.eu/info/law/better-regulation/have-your-say/initiatives/12567-Sustainable-products-initiative_en). For this purpose, IoT and RFID technologies are used to make supply chain management and logistics processes more transparent and efficient. Blockchain technology in particular provides transparency in the supply chain and supports sustainable fashion. In addition to the support it provides for sustainability, digital transformation also provides benefits such as operational efficiency and cost savings, improving customer experience, reaching the market quickly and gaining competitive power, multichannel integration and customer behavior analysis.

As a result of global developments and issues, the textile, apparel, and fashion industries have undergone significant transformations, and it is clear that sustainability and digitalization have emerged as key focal points. On the other hand, an important turning point that accelerated the transformation process was the COVID-19 pandemic that affected the entire world. The pandemic affected supply and demand dynamics in the short and long term. In the short term, businesses stopped their activities, production facilities were closed, orders were canceled and import and export were restricted due to the closure of border gates. Consumers reduced their spending due to economic uncertainty and restrictions. The textile, apparel, automobile, and electronics sectors in particular have suffered badly. Small and medium enterprises struggled with limited resources. The pandemic's long-term effect demonstrated that there has been a call for crisis management and greater resilience in business by identifying vulnerabilities within traditional supply chains and focusing on the need for resilient, responsive, and technologically savvy business models. Companies have turned to strategies such as local sourcing and multi-sourcing to achieve their production and supply chains becoming agile. Digital transformation and e-commerce have been more robust because the pandemic has highlighted the worth of digital technologies in remote work, e-commerce expansion, and virtual try-on experience, and concurrently has emphasized the importance of sustainable practices in a bid to reduce disruption and waste in an ever-more uncertain global economy. The issue of sustainability, which was already a priority on the agenda of manufacturing companies, has been further motivated and stimulated by government policies and regulations. On top of this, the pandemic of COVID-19 impacted human attitudes towards health, the economy, and the environment, eventually determining lifestyles and consumption patterns globally. In particular, concerns about the environmental damage caused by fashion, recycling awareness, and organic and vegan products consciousness have increased, thus sustainable consumption practices have gained momentum. Thus, from design to the post-consumer life, companies are trying to gain a competitive advantage with a sustainable and digital framework. Companies in textile and apparel industries are now adopting a twofold approach, i.e., profitability and sustainability. That is, even collaborations between brands in logistics activities are being brought into the attention, or they collaborate on green initiatives and work together to revamp manufacturing units collectively or construct recycling units (McKinsey, The State of Fashion, 2025).

On the production side of sustainable and digital transformation in the textile and apparel industries, the adoption of sustainable materials has increased to appear like recycled polyester, organic cotton, hemp, and biodegradable materials. These sustainable materials can help lessen reliance on virgin resources and pollution associated with conventional systems of textile and apparel production. Considerations for waste reduction are further enhanced through systems of closed-loop production resulting from principles of circular economy, in which used garments can not only be diverted from landfills but also brought back into the marketplace through processes of collection, recycling, and reusing. Mechanical and chemical recycling help facilitate fiber reuse without compromising quality. Water-efficient dyeing and finishing techniques are also being applied to dramatically reduce water use and wastewater contamination. At the same time, carbon emissions are reduced through the use of renewable energy, reorganization of logistics, and efficient production techniques. While the circular economy model supports the reuse of resources, sustainable supply chain management aims to integrate environmental, social, and economic goals into business decisions. Life Cycle Assessment (LCA) is used for waste management and eco-design by measuring the environmental impacts of products.

Alongside sustainability, digitalization is revolutionizing the textile, apparel, and fashion industries' impact on the environment. By leveraging 3D digital fashion design software and virtual prototyping technologies, fashion companies are reducing the need for physical samples, and waste in textiles, and facilitating a more efficient product development process. Companies are also increasingly hosting virtual fashion shows and digital showroom experiences that resulted in reduced travel emissions and resource usage associated with traditional shows. Furthermore, the emergence of AI-based trend forecasting and demand prediction algorithms is enabling brands to produce more accurately, thus reducing overproduction and excess inventory. QR codes, blockchain-enabled supply chain tracking systems, and smart labels, in particular, give consumers correct information on the processes that produced the item. Hence, a consumer can access a garment's material, factory of production, and carbon footprint, instantly.

On the fashion retailing, consumption habits, and consumer behavior side, increasing environmental awareness and developing digital technologies have caused radical changes in consumer behavior. The fashion industry, in particular, has been one of the areas that has felt the effects of both digital transformation and sustainable consumption most intensely.

The environmental and ethical problems created by fast fashion are increasing the interest in the concept of sustainable fashion day by day. Consumers demand the use of environmentally friendly materials in production processes, the establishment of fair labor conditions and the adoption of transparent supply chain policies. At the same time, a conscious consumer group is emerging that follows the steps taken by brands in sustainable fashion more closely. Environmentally and socially conscious consumers now care not only about the product-related attributes of fashion items but also about the environmental impacts of the production process and the sustainability policies of brands. In particular, products made from recycled materials, and sustainable and ethical production processes have become important elements that shape consumer preferences. This change encourages brands to be more transparent and to use certificates and traceability systems that prove their environmental and social responsibility. Additionally, the circular economy, collaborative consumption, mindful consumption and slow fashion movements help to spread sustainable consumption models.

In addition, digital marketing and online shopping are radically reshaping consumers' fashion consumption habits. The spread of e-commerce platforms, influencer marketing on social media and personalized shopping experiences directly affect consumers' purchasing decisions. Social commerce applications that offer direct shopping opportunities through social media platforms such as Instagram,

TikTok, and Facebook make it easier for brands to reach consumers. The conveniences offered by online shopping increase access to sustainable fashion and offer consumers the opportunity to make more conscious choices. Thanks to digital technologies, brands analyze consumer preferences to offer personalized recommendations that further improve the shopping experience by tailoring product suggestions. In this way, customer service is improved and a faster and more effective shopping experience is provided. Smart labels and QR codes provide sustainability details, while AR and VR technologies enable virtual try-ons, reducing returns and promoting sustainability.

Sustainable fashion is an approach that aims to transform today's consumption habits into an environmentally friendly and ethical structure. However, this transformation process must be constantly examined due to the complex mental structure of consumers and the rapidly changing dynamics of the fashion industry. Consumers' attitudes toward sustainable fashion are influenced by many variables such as demographic characteristics, social influences, cultural differences, and technology. This is actually related to the product attributes and consumer-related motivations in product choices and preferences. Design, fit, brand name, sustainability, and trendiness of the products have varying priorities for consumers in their clothing choices. Consumers may engage in sustainable consumption behaviors such as donating, reusing, and recycling clothing even if they don't purchase sustainable products. In addition to product-related factors already mentioned, consumer-related motivations and drivers are a factor in sustainable fashion consumption. These consumer-related factors are largely due to pro-environmental attitudes (including knowledge, awareness, and concern about the environment) and societal missions. However, due to the dynamic nature of the fashion industry and the impacts of major macro events (such as COVID-19), other motivations are likely to become important in terms of consumer behavior and analysis over time. Meanwhile, due to the influence of social media and digital platforms, consumers' awareness of sustainable fashion has increased, but the extent to which consumer awareness is reflected in purchasing behavior should be monitored.

Cultural differences also play an important role in consumers' perceptions and purchase intentions towards sustainable fashion products. For example, studies have shown that information quality and trust are critical to the slow fashion adoption process in individualist societies such as Canada and South Africa, but these factors do not constitute a significant obstacle in collectivist cultures such as China (Duh et al., 2024). On the other hand, it has been emphasized that social effects should be given priority in collectivist cultures (Su et al., 2023). Moreover, it has been pointed out that sustainable fashion brands should increase perceived value in Asian and North American markets while focusing on aesthetic features and environmental awareness in European markets (Brandão & da Costa, 2021). As a result, brands that want to be successful on a global scale in the sustainable fashion industry need to constantly update their strategies. Only in this way can sustainable fashion be adopted by the masses and a permanent transformation in the fashion industry be achieved.

In line with sustainability goals, it is important for institutions, governments, and brands to provide consumers with information on sustainability in terms of consumer preferences. While conveying correct information and transparent communication positively affect sustainable consumption behavior, lack of trust and concerns about greenwashing are among the factors that negatively affect consumers' purchasing decisions. Therefore, establishing transparent communication, publishing sustainability reports, and building trust through independent audits are required to support sustainable fashion consumption. Social marketing stands out as an effective tool to inform and communicate with consumers transparently. Sustainable consumption and social marketing work together to reduce the environmental impact of fashion and improve its ethical practices. Social marketing can also play a role in achieving social change as a possible approach to eliminate the attitude-intention-behavior gap in sustainable consumption behavior. As part of their social marketing activities, brands adopt a marketing approach that develops strategies to change consumer behavior through practices such as awareness and educational campaigns, encouraging value-oriented consumer behaviors, communication on ethics and transparency, encouraging participation in sustainable fashion trends, and incentives encouraging behavior change. Brands create awareness by organizing campaigns on topics such as recycled fabrics, ethical production processes and reducing carbon footprint. They tell consumers about the social and environmental impacts of making sustainable choices. In this way, sustainable consumption habits such as purchasing higher quality products, buying second-hand clothes, repairing clothes, renting, and recycling are encouraged. Social marketing also ensures that brands adhere to ethical values in their production processes and follow a transparent communication strategy. Social marketing uses social media campaigns and influencer collaborations to support sustainable fashion movements. Brands use reward systems to encourage consumers to adopt sustainable consumption habits. Giving discount coupons to consumers who return old clothes is an example of an incentive strategy in social marketing.

Advertisements can also be used to persuade consumers to choose sustainable fashion products. Within the scope of social marketing, frequently used negative emotional appeals (fear, guilt, and shame) are generally used by nonprofit organizations and in areas that concern public health. This type of advertising can be used to persuade consumers to avoid excessive consumption or to turn to sustainable alternatives in fashion consumption. From that point of view, consumer insight provides a comprehensive understanding of consumers' attitudes toward sustainable fashion through social marketing. However, consumer research in this area is very limited and almost nonexistent.

Another factor that transforms consumer behavior in fashion retailing is developing technologies and digitalization. Digitalization transforms the consumer experience and makes fashion shopping easier. E-commerce offers an opportunity to discover and purchase sustainable products without even going to physical stores. Through mobile commerce growth, phones have now become the shopping platform. 92.6% of global internet users actively use mobile applications (Pop et al., 2023). Reasons to utilize mobile shopping applications are based on utilitarian and hedonic reasons and social influence. Utilitarian reasons comprise time and money saving, easy access to information, product variety and customization, easy-to-use interface and productivity-enabling features, while hedonic reasons involve entertainment, discovery and new ideas, bargain hunting, impulsive shopping inclinations. Social reasons involve family, friends and peer influence, influencers, and social media suggestions. Although consumers from different generations accept mobile shopping for different reasons, it is of great importance to develop and implement strategies that will have a positive impact on mobile shopping experience and preferences.

Along with the wide spread of e-commerce, digital transformation has opened the doors to a new era for consumers in fashion retailing. Modern technologies and data science are at the heart of the retail industry. These tools provide both direct solutions and enhance the experience at customer touchpoints. Fashion brands use these technologies to improve their operations, increase service quality, personalize the products and services according to customer needs, and deliver a remarkable shopping experience. The use of modern tools within the scope of retail operations includes the use of Generative AI in new product designs, verification of product authenticity with computer vision and machine learning against counterfeiting, inventory management, trend forecasting, transparency in the supply chain and resource tracking. The use of modern tools to improve service quality in fashion retailing includes smart store experience, in-store customer movement analysis, visual product search technologies, and various payment tools (such as NFC, crypto money, QR code, chatbot, and biometric payment systems).

The use of digital technologies and tools in fashion retailing has given birth to smart retailing as a concept. Retail stores are equipped with technologies such as AI, big data, IoT, blockchain, RFID, QR codes, AR, and VR for enhancing the customer experience and making businesses more efficient. With self-service checkout, contactless payments, and AI-assisted shopping assistance, the shopping experience at stores has become faster and more efficient. Smart shelves and electronic labels maximize inventory management while providing customers with more data. On the e-commerce side, with AI-powered chatbots and customer services offering personalized recommendations, customers can make better decisions while shopping. AI optimizes inventory management, improves price strategies, and offers personalized recommendations based on previous shopping trends. Consumer behavior is better understood with big data analytics, which also guides marketing strategies. While IoT and RFID technologies enhance tracking consumers and inventory management, blockchain is used in supply chain management and secure payment systems.

In fashion retail, augmented reality (AR) and virtual reality (VR) technologies are changing the physical store experience through new innovations such as virtual fitting rooms and smart mirrors. A variety of brands are embedding AR and VR technologies into mobile applications to improve interactivity and personalization for consumers. These tools enable customers to try clothing and accessories in a digital format. In addition, automated AI chatbots and smart assistants assist and improve the shopping experience through improved automation of customer service. In particular, virtual fitting rooms and smart mirrors give shoppers the ability to see themselves in clothing without needing to go to a physical store, while providing better customer satisfaction and reducing returns due to selecting the wrong size, assisting consumers in purchasing more responsibly, and contributing to greater sustainability. Contactless payments and digital wallets enhance security and speed related to the shopping experience. The metaverse also offers opportunities for brands to open stores and sell digital products within virtual worlds. Eventually, smart retail technologies increase customer satisfaction, reduce costs for businesses, and provide a competitive advantage. These innovative technologies are transforming the customer experience in fashion retail however, traditional retail elements have not completely disappeared and their use in a balanced way together with modern tools is essential. In this context, studies that evaluate consumers' approaches to traditional and modern retail tools together are of great importance. In today's retail environment, especially where e-commerce is growing, it is critical for brands to analyze how tools are positioned together and how they impact the consumer experience when shaping their strategies. The integration of traditional and modern tools provides valuable insights into the shopping habits of different sociodemographic groups, contributing to more informed and targeted decisions in the retail sector.

Thus, sustainability and digitalization on the consumer side are radically changing shopping habits. Digital markets and e-commerce applications in fashion retail make shopping processes more efficient and personalized, while also facilitating access to sustainable products. Digital technologies such as artificial intelligence-supported recommendation systems, virtual try-on opportunities with AR, and data analytics-based stock management offer customers a more conscious and faster shopping experience, while also providing operational efficiency to businesses. All of these show that digitalization in the retail sector brings not only environmentally friendly solutions but also innovations that increase competitiveness and customer satisfaction. To keep up with this transformation, brands are shaping the retail world of the future by investing in both sustainable production policies and innovative digital solutions.

Through the incorporation of these innovations, the textile and apparel industries are moving towards a more sustainable and technology-oriented future. However, there are challenges related to scalability, cost, and changes in consumer attitudes that must continue to be resolved through continuous stakeholder engagement, technological innovation, and policy support in order to achieve long-term goals. Therefore, precisely at that point, the analysis of the sustainable consumption pattern and reasons of consumers and the assessment of the impact of COVID-19 can make brands learn about the consumer and develop appropriate strategies to become successful. Understanding how post-pandemic dynamics shape consumer decision-making and purchasing behavior in the textile, apparel and fashion industry by analyzing changes in consumer behavior and consumption of fashion products will be beneficial for both researchers and practitioners. Fashion products are retailed through various channels and mediums with varied marketing strategies in various marketing mixes, and the changes in consumer behavior must be monitored continuously through longitudinal studies. Fashion consumption may differ by various consumer groups and geographic places. An investigation of all the above dimensions of fashion consumption provides several managerial implications for professionals in the industry.

To briefly mention the contents of the following chapters of this book; the second chapter entitled "Sustainable Fashion Consumption in a Transformative Era" starts with an introduction to the sustainability concept. In this respect, a historical background of sustainability, from its origins to the laws, rules, and legislation, is described, followed by the relevant sustainability concepts in fashion supply chain, marketing, and retailing. The adoption of sustainability in fashion supply chains and retail businesses is explained with numerous recent examples and cases, including some practices in regard to the circular economy and blockchain operations. This part ends with the discussion about the impact of the COVID-19 pandemic on the producers', brands', and consumers' approaches to sustainability, highlighting the heightened sensitivities about environmental concerns. In the following part, the consumer behavior and consumption patterns are provided in three aspects, which cover the consumption of sustainable fashion products in general, preference for sustainable fashion products produced by using environmentally friendly materials and processes, and preference for sustainable fashion products with the motivation of decreasing consumption via collaboration consumption-based models. At the end of this part, the approaches of different consumer groups (belonging to different cultures, genders, ages, and generations) towards sustainability concepts and sustainable fashion products are brought up. Putting special emphasis on the recent studies, practical insights for the practitioners are summarized at the end of each sub-part. In the last part, an empirical study examining sustainable fashion consumption among Generations Y and Z by developing a conceptual framework that integrates individual beliefs (self-interest and pro-social motives) with external influences is presented, and its results are given. The framework combines the Norm Activation Model (NAM) and Theory of Planned Behavior (TPB), incorporating the effects of pandemic-induced awareness and social responsibility. A multigroup analysis compares generational differences, and the findings are used to provide recommendations for both academics and practitioners.

In the third chapter entitled "Digital Markets and E-Commerce Applications in Fashion Retailing", a broad overview of the digital fashion retail and e-commerce market, the features of e-commerce applications including company websites, third-party marketplaces and mobile applications in fashion retail are offered, and consumer tendencies are presented, specifically for mobile shopping applications, the use of which is rapidly increasing. The chapter starts with an overview of the digital fashion market at global and regional levels, including market size, applications, dynamics, and consumer trends with current statistical indicators. It then presents a comprehensive portrait of e-commerce applications in fashion retail, encompassing online and mobile shopping. The chapter then discusses consumer characteristics, motivation, and adoption of mobile shopping applications, which have become more widespread, especially as technology becomes more prevalent in people's lives. Understanding the priorities and the expectations of the consumers is beneficial particularly for the marketers to leverage their sales and gain competitive advantage in online fashion retail. The empirical study at the end of the chapter investigates the adoption of mobile shopping applications by an integrated model based on TAM (Technology Acceptance Model) and TPB (Theory of Planned Behavior) within the framework of the S-O-R (Stimuli-Organism-Response) algorithm extended with a particular factor, trust, to analyze consumer behavior regarding the adoption and future use of mobile shopping applications for purchasing apparel products. To understand the extent to which the pandemic, which has significantly affected the transformation process towards digitalization in retail, has affected consumers, the results are validated by applying a multigroup analysis among two groups of participants affected at different levels by the pandemic. The model and the analysis of the empirical study are expected to contribute to the academics studying in that field, and the findings may serve marketers and retailers in realizing the points to be focused on in terms of innovative contents that will reach a wider range of consumers.

The fourth chapter entitled "The Role of Social Marketing on Promoting Sustainable Fashion Consumption", presents an overview of sustainability in fashion and the concept of social marketing as an innovative approach to promoting sustainable consumption, with examples from brands and addressing its main characteristics. In addition, sustainable consumption behavior will be examined within the framework of social marketing appeals and self-benefit and other-benefit motivations of consumers. The chapter starts with the concept of sustainability in fashion with production, consumption, and marketing dynamics. Then, as a way of persuading consumers to question their consumption expenditures and encourage sustainable consumption, social marketing with its key features and activities of fashion brands in this regard is presented. Any reference book is not available in the literature covering the works of brands in that field and the use of social marketing to encourage sustainable consumption in fashion. In this respect, it is considered to be a resource for both industrial practitioners and academic researchers. At the end of the chapter, a field study for investigating the consumers' behavior toward sustainable fashion consumption searching the impacts of social marketing appeals (fear, guilt, and shame) and consumer identity appeals (self-benefit, and other-benefit) is conducted. Many brands are now trying to encourage consumers to sustainable fashion consumption through social marketing campaigns. At this point, the findings can serve as a guide for maximizing the effects of social marketing activities.

In the fifth chapter, entitled "Modern Tools in Fashion Retailing and Consumer Perspective," the modern tools and technology used in fashion and retail businesses and the consumers' approaches towards them are presented. Firstly, the modern technology (artificial intelligence, computer vision, big data analytics and the internet of things, robotics) related to data science lying beyond the implementation of modern tools is introduced with the explanations about their potential use in retailing. Then the modern tools and technology are considered from the retailers' perspective, concentrating on the modern tools at customer touchpoints (such as personal recommendation systems, chatbots, smart mirrors, virtual try-ons, novel payment methods). The retailer's perspective is examined in four aspects of benefits in the improvement of retail operations, service quality, mass customization, and consumer experience by giving some examples and cases from the fashion market, which can be helpful for practitioners. A special emphasis is given to virtual reality and augmented reality applications because of their broader applications and specific use for fashion marketing and selling. Secondly, the consumer's approach to these tools in both smart and fashion retailing is investigated by focusing on the motivations, expectations, and consumer characteristics and summarizing practical implications. Thirdly, the findings of an empirical study whose aim is to analyze the service quality perception of both modern tools (including smart assistants and virtual try-ons) and traditional tools (servicescape in mobile shopping) in the enhancement of store atmospherics and experience in mobile shopping applications for fashion retailing companies are given. In the empirical study, the impact of service quality perception of modern tools on consumers' cognitions and emotions is compared with that of the traditional tools. The results of the study are evaluated according to different sociodemographic characteristics such as gender, age, nationality, and the experience level of the consumers by providing some practical advice to the practitioners.

References

Brandão, A., & da Costa, A. G. (2021). Extending the theory of planned behaviour to understand the effects of barriers towards sustainable fashion consumption. *European Business Review, 33*(5), 742–774.

Correia, M. S. (2019). Sustainability: An overview of the triple bottom line and sustainability implementation. *International Journal of Strategic Engineering (IJoSE), 2*(1), 29–38.

Duh, H. I., Yu, H., Venter de Villiers, M., Steffek, V., & Shao, D. (2024). Determinants of young adults' slow fashion attitudes and idea adoption intentions in Canada, China and South Africa. *International Journal of Emerging Markets*.

https://commission.europa.eu/energy-climate-change-environment/standards-tools-and-labels/products-labelling-rules-and-requirements/ecodesign-sustainable-products-regulation_en (accessed 13 Mar 2025a).

https://commission.europa.eu/strategy-and-policy/priorities-2019-2024/european-green-deal_en (accessed 24 Apr 2025a).

https://ec.europa.eu/info/law/better-regulation/have-your-say/initiatives/12567-Sustainable-products-initiative_en (accessed 24 Apr 2025b).

https://environment.ec.europa.eu/strategy/circular-economy-action-plan_en (accessed 24 Apr 2025c).

https://environment.ec.europa.eu/strategy/textiles-strategy_en#related-strategies (accessed 24 Apr 2025d).

https://sdgs.un.org/goals (accessed 13 Mar 2025b).

https://unfccc.int/sites/default/files/resource/US_accessibleLTS2021.pdf (accessed 24 Apr 2025e).

https://www.epa.gov/circulareconomy (accessed 24 Mar 2025f).

https://www.fashiondive.com/news/US-state-legislatures-fashion-sustainability-safety/738813/ (accessed 14 Mar 2025).

https://www.oecd.org/en/topics/extended-producer-responsibility-and-economic-instruments.html (accessed 23 Mar 2025).

McKinsey, The State of Fashion 2025., https://www.mckinsey.com/industries/retail/our-insights/state-of-fashion (accessed 24 Apr 2025).

Pop, R. A., Hlédik, E., & Dabija, D. C. (2023). Predicting consumers' purchase intention through fast fashion mobile apps: The mediating role of attitude and the moderating role of COVID-19. *Technological Forecasting and Social Change, 186*, 122111.

Su, J., Iqbal, M. A., Haque, F., & Akter, M. M. K. (2023). Sustainable apparel: A perspective from Bangladesh's young consumers. *Social Responsibility Journal, 19*(9), 1654–1670.

Sustainable Fashion Consumption in a Transformative Era

Keywords Historical background for sustainability · Sustainability concepts · Sustainability practices · Fashion supply chain · Fashion retailing · Consumer behavior · Theory of Planned Behavior (TPB) · Norm Activation Model (NAM) · COVID-19

1 Introduction

The fashion industry has experienced rapid growth in the past few decades, achieving a total value of US$3 trillion and accounting for 2% of the world's total gross domestic product (Chen, 2024). However, the fashion industry is the world's second most polluting industry, responsible for 20% of wastewater (79 billion cubic meters of water), 10% of carbon emissions (1.7 billion tons of CO_2 emissions), and 92 million tons of textile waste annually (Chen, 2024; Centobelli et al., 2022). If the fashion industry continues with the current practices worldwide, the carbon emissions are expected to reach one-quarter in 2050, the material consumption is estimated to be 167 billion tons in 2060, and the input of fossil fuel feedstock is forecasted to be 160 million tons by 2050 (Bhandari et al., 2022; Dissanayake & Weerasinghe, 2021).

The fashion industry generates pollution at every stage of the production process. The damage done to the environment starts with the fiber production stage. Natural fiber production requires a high amount of water, while synthetic fiber production results in a high carbon footprint and high energy requirements (Niinimäki et al., 2020). Environmental pollution is enhanced with the chemicals used in the production stages. Garment production generates a large quantity of textile waste. 15% of fabric becomes pre-consumer waste in garment production, and one-third of products are not sold at all because of overproduction (Niinimäki et al., 2020).

The excessive waste is also due to consumption, which has grown significantly with the adoption of fast fashion and online retail. Offering trendy and affordable garments, fast fashion has shortened the apparel life cycle and encouraged a higher

C. Saricam, N. Okur, *Changing Textile and Apparel Consumption in Transformative Era of Sustainability and Digitalization*, Sustainable Textiles: Production, Processing, Manufacturing & Chemistry,
https://doi.org/10.1007/978-3-031-97419-9_2

frequency of purchasing (Cai & Choi, 2020; Wang et al., 2020). The prices of the apparel products have decreased by more than 30% between the years 1996 and 2018, but the apparel product purchase increased by 40% in Europe over the period from 1996 to 2012 (Dissanayake & Weerasinghe, 2021). Online fashion retailing worsened the consumption by enhancing the convenience of shopping. The amount of clothing production doubled in 15 years between the years 2000 and 2015, and global consumption is projected to reach 102 million tons by 2030 (Dissanayake & Weerasinghe, 2021; Niinimäki et al., 2020). Depending on this huge amount of consumption, 5 million tons of clothing are discarded, either incinerated or sent to landfills, every year solely in European Union countries (*EU Strategy for Sustainable and Circular Textiles*, 2024).

The environmental damage and massive consumption trend have been addressed seriously by various organizations and bodies such as the United Nations, the European Union, and governmental institutions all around the world. This has led to the creation of many regulations, policies, and frameworks with the purpose of reducing environmental harm and providing social welfare. Parallelly, the prominent, global fast-fashion brands recognized the importance of sustainability, and they showed their commitment to addressing environmental and social challenges by setting ambitious climate goals (Karaosman & Marshall, 2023). The branded retailers and the manufacturing companies began to implement ecological practices in their production to reduce waste and environmental consequences (Kumar et al., 2021).

The brands turned to the issues of sustainability to satisfy the expectations of their consumer base. In fact, consumers are also becoming much more environmentally conscious, and they are shifting towards sustainable consumption patterns (Tewari et al., 2022; Leclercq-Machado et al., 2022b). The increasing environmental awareness and sense of social responsibility resulted in more sustainable behavior (Floriano & Matos, 2022). Besides, the circular business and collaborative consumption-based models having the primary purposes of lengthening the useful life and decreasing the spending on apparel products are gaining more attention from consumers every day. The consciousness and sensitivity level of the specific group of consumers are higher. Explicitly, Generation Z has a higher level of awareness and knowledge of both ecological and social problems, and it is more motivated to exhibit pro-environmental behavior and switch to sustainable products (Abrar et al., 2021).

On the other hand, the Covid-19 pandemic also became influential in the fashion market in regard to the awareness of sustainability, consumption patterns, and shopping habits. The health concerns and the anxiety felt during the Covid-19 pandemic transformed into regret for the harm done to the environment and therefore played an important role in the purchasing of eco-friendly products (Kumar et al., 2021). People have begun to appreciate the social and environmental issues during the Covid-19 pandemic as a response to these upsetting conditions (Alexa et al., 2021). Therefore, the Covid-19 pandemic became a starting point and prompted consumers to consider the environmental impacts of their fashion procurements and to adopt eco-friendly products more (Sobuj et al., 2021).

The Covid-19 pandemic also became a turning point for consumers to change shopping habits. The consumers began to demand online purchasing more because of health concerns. Even though the major influence of the Covid-19 pandemic has disappeared, the new shopping habits gained during the Covid-19 pandemic have become permanent.

Furthermore, the Covid-19 pandemic also triggered a digital shift in fashion marketing and retail operations by enhancing digital transformation and dematerializing the resource-intensive practices in traditional supply chains (Casciani et al., 2022). The digital transformation due to the advancements in data science was accelerated because of the obligations that have arisen in the Covid-19 pandemic. Therefore, the Covid-19 pandemic became a turning point in the transformative era described with the substantial changes in terms of sustainability and digitalization.

Considering all the points outlined above, this chapter examines sustainability in the fashion industry from the perspectives of production, retailing, and consumption within the transformative era. Specifically, the following section begins with some explanations about the relevant policies and regulations, followed by the definitions of key sustainability concepts in textile, apparel, and fashion retail businesses. The section proceeds to explore the sustainability practices in the fashion supply chain, providing numerous examples from the fashion brands and textile companies, and concludes by highlighting the role of the Covid-19 pandemic as a pivotal turning point in the industry's transformation process. The subsequent section examines consumers' perspectives on sustainability, focusing on sustainable fashion product categories. Each subsection concludes with the practical implications suggested by the researchers. Finally, the chapter ends with the presentation of results of an empirical study whose aim is to analyze the behavioral intention of different generations towards sustainable fashion products by pointing out the impact of the Covid-19 pandemic on behavioral intentions.

2 Sustainability in Fashion Supply Chain and Fashion Retail

2.1 Sustainability Concepts in the Textile, Apparel, and Fashion Retail Businesses

Sustainability is a concept that has evolved over time with the positive contribution of many stakeholders, including the governments, organizations, companies, and brands. Therefore, it has many aspects associated with the governmental strategies and the current applications on both the supply and market sides. In this context, this section summarizes the major international strategies and initiatives in regard to sustainability, followed by the descriptions of the current sustainability concepts in supply chain, marketing, and retailing.

2.1.1 Historical Background for Sustainability and Relevant Regulations

The concept of sustainability came into existence in the second half of the twentieth century. The Stockholm Environment Conference, established by the United Nations (UN) in 1972, became the first platform enabling the countries to discuss the global environmental problems and to cooperate on solutions. The Brundtland report, namely "Our Common Future," published by the World Commission on Environment and Development in 1987, became a milestone in sustainable development and sustainability efforts (Thorisdottir & Johannsdottir, 2020). The Brundtland Commission made the most well-known definition of sustainable development as "*development that meets the needs of the present generations without compromising the ability of the future generations to meet their own needs*" (Alhaddi, 2015). With this definition, sustainability has assigned a responsibility to every individual and supported the belief that the world and scarce resources are already inherited from future generations. The report also outlined four key principles to achieve sustainability as follows: focusing on the responsible management of natural resources, emphasizing the use of environmentally friendly materials, stressing the promotion of practices in an efficient ecosystem, and advocating the contribution of society, ensuring that the sustainable practices are accessible by all people (Thorisdottir & Johannsdottir, 2020). The Rio Earth Summit, established in 1992, became the meeting where the global documents "Rio Declaration" and "Agenda 21" were accepted. In the same meeting, the UNFCCC- United Nations Framework Convention on Climate Change—was signed by the member countries, which continued with the Kyoto Protocol in 1997 (the first international climate agreement aiming to reduce greenhouse gas emissions) and the Paris Climate Agreement in 2015 (the first international agreement aiming to fight global warming and climate change). In the same year as the Paris Climate Agreement, the UN General Assembly accepted the Sustainable Development Goals (SDGs) as a part of the "2030 Agenda for Sustainable Development," calling on everyone, including governments, companies, civil organizations, and the public, to contribute to the achievement of the goals (Cai & Choi, 2020). 17 SDGs were listed as SDG1: No Poverty, SDG2: Zero Hunger, SDG3: Good Health and Well-being, SDG4: Quality Education, SD5: Gender Equality, SDG6: Clean Water and Sanitation, SDG7: Affordable and Clean Energy, SDG8: Decent Work and Economic Growth, SDG9: Industry, Innovation, and Infrastructure, SDG10: Reduced Inequalities, SDG11: Sustainable Cities and Communities, SDG12: Responsible Consumption and Production, SDG13: Climate Action, SDG14: Life below Water, SDG15: Life on Land, SDG16: Peace, Justice, and Strong Institutions and SDG17: Partnerships for the Goals. The SDGs provided measurable guidance in alignment with the Triple Bottom Line (TBL) approach, which defined three pillars of sustainability as planet (environment), people (social), and profit (economic) (Fung et al., 2020). Even though the international agreements and the principles set by the UN are based on volunteerism, many countries integrated them into national and international policies and regulations.

In fact, the European Union (EU) has emerged as a leading authority on environmental policies. The policy development and regulations gained acceleration in parallel with the Brundtland report. Some milestones became the acceptance of European Environmental Policy (2001), European Climate Change Strategy (2005), and European Climate Change and Renewable Energy Package (2007). With this package, the EU has set targets for 2020, including the reduction of greenhouse gas emissions by 20% compared to 1990 levels, the increase of the share of renewable energy use to 20%, and the improvement of energy efficiency by 20%. The EU Commission accepted the "7th Environment Action Programme," which aimed to reach the SDGs in 2008. The EU has continued to develop policies such as "Europe 2020: A European Strategy for Smart, Sustainable, and Inclusive Growth (2010)," "EU Biodiversity Strategy to 2020 (2011)," "Roadmap to a Resource Efficient Europe (2011)," "An EU Action Plan for the Circular Economy (2015)," "A European Strategy for Plastics in a Circular Economy (2018)," and "The European Green Deal (2020)" (Eckert & Kovalevska, 2021). The Green Deal has become an integral part of the strategy developed by the current Commission to implement the Sustainable Development Goals and provide a set of policy initiatives with the purpose of reducing the net greenhouse gas emissions by at least 55% by 2030 and making the EU climate neutral in 2050 (Szpilko & Ejdys, 2022). The European Commission emphasized the contribution of all sectors, such as construction, energy, transport, agriculture, food, and textiles. In this respect, the "EU Strategy for Sustainable and Circular Textiles" was set in March 2022, which has aimed to coordinate the transition actions towards more sustainable and circular production, consumption, and business models in the textile industry (Centobelli et al., 2022). The EU Strategy implied that the textile goods sold in the EU should have been composed of mainly recycled fibers and free of harmful compounds. Moreover, they should have been manufactured to have long-lasting and recyclable characteristics according to the social and environmental standards by 2030. Besides, it was stated that fast fashion needs to become outdated (European Parliament, n.d.). Based on the EU strategy, a series of regulations, legislative measures, and other activities have been proposed or planned to be implemented. Specifically, the "Ecodesign for Sustainable Products Regulation" has been established, which defines the requirements for more durable, easy-to-repair, and recyclable products. The regulations for ecodesign have also involved the measures about Digital Product Passport (DPP), which has required the enterprises to present detailed information on the sustainability and recyclability of products. The "Zero Pollution Action Plan" has been established in order to reduce the release of microplastics, and the "Waste Shipment Regulation" has restricted the export of textile waste. The mandatory and harmonized "Extended Producer Responsibility (EPR) Rules" have been introduced in order to hold manufacturers and importers responsible for reusing and recycling the worn clothing. Finally, the "Textile Labelling Regulation" has been revised, and the "Empowering Consumers in the Green Transition Directive and Green Claims Directive" has been proposed in order to increase the awareness of consumers and prevent greenwashing (EU Strategy for Sustainable & Circular Textiles, 2024).

In fact, not only has the EU Commission taken the initiatives as mentioned above, but almost all the countries in the world have integrated the sustainable development goals into their policies and legislations. For instance, China also adopted a circular economy to struggle with the environmental pollution and the scarcity of resources. Canada introduced Extended Producer Responsibility Schemes (EPR) and a number of zero-waste programs in order to promote a circular economy (Dissanayake & Weerasinghe, 2021). However, the policies developed by the other countries than the EU did not present a holistic approach, as in the case of the EU.

On the other hand, the national/international joint initiatives or nongovernmental organizations, as well as the global brands, also have adopted the sustainable development goals and set themselves targets for sustainable development, as the fashion companies are expected to commit to corporate social responsibility, promote more sustainable production and consumption models, reduce water consumption, and ensure decent working conditions (Abbate et al., 2024). For instance, H&M has made a commitment to reduce carbon emissions by 50% by 2030 and reach net zero emissions by 2040. Inditex has also committed to transforming its supply chain, extending its circularity projects, and helping to improve biodiversity. Moreover, almost any company or brand establishes many strategies and action plans in order to improve sustainability.

2.1.2 Sustainability Concepts Related to Supply Chain

Sustainability and sustainable development are possible with the implementation of special strategies, models, and practices. An inevitable strategy to be applied is the transformation of traditional supply chains into sustainable supply chains. The difference between the traditional supply chain and the sustainable supply chain is that while the traditional supply chain aims to achieve long-term performance in terms of speed, cost, and reliability of operations related to the management of raw materials, inventories, manufacturing, and finished goods for both the individual companies and the supply chain as a whole, the sustainable supply chain integrates the triple bottom line—environmental, social, and economic goals—into business decisions (Oguntegbe et al., 2023; Sarasi et al., 2023). The environmental goals are achieved with the creation of ecological value and reduction of consumption, whereas the social goals are achieved with the implementation of ethical labor practices and the promotion of equality. Finally, the economic goals are tried to be achieved with the financial profitability and growth (Arrigo, 2020; Daukantienė, 2023). Therefore, a competitive advantage is obtained while ensuring at least a neutral effect on the natural and social environment in sustainable supply chains (Garcia-Torres et al., 2022).

Nonetheless, the establishment of a sustainable supply chain is not sufficient on its own. Therefore, most of the companies prefer to implement a broader model, the circular economy. The circular economy is defined as "*an economic system that replaces the 'end-of-life' concept with reducing, alternatively reusing, recycling, and recovering materials in production/distribution and consumption processes.*" (Wang et al., 2020). In a circular economy, it is aimed to reduce the consumption of resources and to introduce the post-use materials into the production process again

(Dissanayake & Weerasinghe, 2021). Even though it is not a new concept, the applications based on the circular economy model gained popularity. Because the circular economy model is not specific to production operations, it can be applied to any type of business, as understood from the definition.

The circular economy can more clearly be described with two approaches: Cradle-to-Grave (C2G) and Cradle-to-Cradle (C2C). The C2G approach begins with the resource extraction, includes manufacturing, and ends up with the product in a landfill (Balwan et al., 2022). C2G has an open-loop system and is called a linear (make-use-dispose) model. Conversely, C2C refers to manufacturing, including the design and production of products that can be recycled and sent back to the ecosystem. C2C has a closed-loop system and is therefore defined as a circular model (Fung et al., 2020; Wang et al., 2020). Therefore, the sustainable value chain working on the principles of the circular economy model is expected to extend its operations to cover various reverse value chain operations (Paras & Pal, 2024). All the forward activities, including design, dyeing, production, logistics, distribution, retailing, and consumption, are brought together with the reverse activities, including reuse, remanufacturing, and recycling, in order to close the loop in circular supply chains.

Circular economy models focus on the eco-effective use of resources, keeping resources in use as long as possible, and minimization of waste (even striving for zero waste) (Balwan et al., 2022). Therefore, the circular economy is highly associated with the terms eco-design, design for environment, and zero waste. Eco-design is the management of a product's life cycle by increasing the longevity through the recovery and reuse of materials (Choi & Hwang, 2015; Henninger et al., 2021). Similarly, the Design for Environment supports the circular economy and promotes recycling and circulation (Fung et al., 2020). Nonetheless, there is a slight difference between these two terms. Whereas eco-design involves both the design and production processes, and it has the main principle of reducing the environmental impact by using sustainable materials and efficient processes, the Design for Environment approach is more focused on the environmental impacts. Life Cycle Analysis (LCA), which is a method used to measure the environmental impacts of the manufacturing and production processes of a specific product from raw material collection to disposal, can be beneficial in the design process. In fact, LCA helps to track the environmental impact of a specific product during each stage of the supply chain, allowing the identification of the products and processes with the greatest impact. Therefore, it has the ability to determine the key processes to decrease the environmental impact (Wren, 2022). Another term that is associated with the design and circular economy model is the zero-waste design. The main objective of zero-waste design is producing zero waste at every stage of production. Zero waste design is based on the Zero Waste Hierarchy. The zero-waste hierarchy proposes a prioritized list of initiatives to avoid the disposal of the product while allowing it a longer time in circulation. In the zero-waste hierarchy, the first step avoids creating waste by rethinking the design and refusing unnecessary materials. The second step minimizes waste by using fewer materials and reusing products or materials wherever possible. The third step is recycling materials and composting organic waste to reduce environmental impact. The fourth step extracts valuable materials or chemicals from waste for reuse, and the final step manages any remaining waste responsibly (Brydges et al., 2020).

On the other hand, transparency and traceability have critical importance in a circular economy. Whereas transparency makes the information about materials, factory locations, environmental impact, and recycling visible, traceability helps to map the product's origins and discuss end-of-life options for reduced impact (Dragomir & Dumitru, 2022). These two together have the potential to decrease the waste and increase the efficiency of the recycling process as well as eco-design. Specifically, traceability improves the life cycle analysis, and transparency is important for compliance with regulations, especially the establishment of DPP. The DPP presents a digital collection of information, ensuring the tracking of the product in the life cycle, allowing for the assessment of information about the environmental impact and recycling and reusing processes. Moreover, traceability and, particularly, transparency prevent greenwashing, which refers to a common irresponsible practice in the textile supply chain where consumers are willingly deceived by providing false claims and deceptive information about the environmental friendliness of a product (Oguntegbe et al., 2023). Thus, the deterioration of trust in sustainable practices can be eliminated with transparency and traceability in the long run.

2.1.3 Sustainability Concepts Related to Marketing and Fashion Retail

Sustainability is also expected to be achieved in retail operations, and there are sustainability terms used to express specific types of products and business models.

The most general and inclusive concept in fashion retail is sustainable fashion. Bhandari et al. (2022) describes sustainable fashion as a process that covers sourcing, manufacturing, and consumption of the clothes with the responsible utilization of resources without damaging the environment. Brydges et al. (2020) adds that sustainable fashion stands for all efforts in every stage of the product's life cycle, including design, raw material production, manufacturing, transportation, storage, marketing, and final sale, as well as the use, reuse, repair, remaking, and recycling of products. Defining sustainable fashion as a business model that is seeking profitable and fair business systems, Palomo-Domínguez et al. (2023) express that sustainable fashion is a general term involving the main categories in regard to the materials ("recycled," "organic," or "cruelty-free animal products"), the production process ("hand-craft," "vintage," or "custom-made fashion"), the local production ("locally produced fashion"), and the "fair trade" system. Therefore, sustainable fashion is an umbrella term and encompasses many aspects of fashion products.

Some sustainable fashion terms mainly focus on the selection of fibers and processes. The major objectives are the use of biodegradable, organic, or recycled materials and the reduction of environmental impacts in terms of air and water pollution (Khare et al., 2020). In this respect, "environmentally friendly fashion" corresponds to the use of environmentally friendly materials and processes in production. "Eco-fashion" emphasizes the reduction of environmental impact with the selection of accurate processes and efficient use of resources. Whereas "organic fashion" means enhanced organic content, "biodegradable fashion" signifies the selection and production of environmentally disposable materials. "Green fashion" denotes increasing the recycled and organic material content and decreasing the environmental impact. On the other hand, "vegan fashion" avoids the animal-based products and aims to decrease the environmental damage done to the environment.

The other sustainable fashion terms mainly focus on the social aspects of sustainability. The main emphasis of "ethical fashion" is the application of fair-trade principles and sweatshop-free labor conditions (Mukendi et al., 2020). Similarly, "fair trade fashion" focuses on safe working conditions, fair wages, prevention of child labor, and supporting the local businesses. On the other hand, "regenerative fashion" aims to regenerate nature and improve the ecosystems and supports the farmers as a part of the ecosystem.

Yet the other group of sustainable fashion terms highlights the longevity of products and waste reduction. Slow fashion is characterized by the timeless design, the quality materials, the long-term use depending on the durability of the fabrics, and comparatively higher costs (Palomo-Domínguez et al., 2023). "Circular fashion" addresses the use of circular economy principles in the whole supply and value chain operations, including the postproduction stages. Associated with circularity, "zero waste fashion" focuses on zero-waste design and the reduction of waste, "recycled fashion" aims at the reuse of the products, and "upcycled fashion" stresses the use of waste and discarded material for new and valuable products.

Apart from the sustainable fashion concepts in terms of fashion products, there are also sustainable business models in fashion retailing. Most of these models have their roots in collaborative consumption, which is defined as "*an economic model based on sharing, swapping, trading, or renting products and services, enabling access over ownership*" (Botsman & Rogers, 2010). Nonetheless, collaborative consumption itself is a subcategory of the sharing economy, whose other categories are collaborative economy, on-demand economy, peer-to-peer (P2P) economy, zero-marginal-cost economy, and crowd-based capitalism (Selloni & Selloni, 2017).

Having a broader field of application, the sharing economy is defined as "*an economic model based on sharing underutilized assets from spaces to skills to stuff for monetary or non-monetary benefits.*" (Mukendi et al., 2020). The sharing economy has the main principle of sharing the assets already in use. But while doing this, the sharing economy models also suggest some contribution to the circular economy because the unused capacities, such as clothing, textiles, and accessories, are repurposed by preventing them from ending up in landfills and extending their overall lifecycle (Henninger et al., 2021).

Therefore, the collaborative consumption-based models, including the second-hand and rental platforms, are circular economy business models, which are gaining higher significance nowadays. The second-hand platforms (e.g., ThredUp, Poshmark, and The RealReal from the USA; Depop from the UK; Vinted from Lithuania; and Dolap from Turkey) and rental platforms (Rent the Runway, Hurr Collective from the USA; By Rotation from the UK; and GlamCorner from Australia), which work mostly online as B2C or P2P platforms, are the examples for collaborative consumption-based models well adopted by both retailers and consumers.

The other business model based on sustainability is the Product-Service-System (PSS), which refers to the combination of products and services. PSS covers repairing (The Restory, Sojo, The Repair Cafe, and Clothes Doctor are some examples of repairing services), refurbishing, and upcycling in order to increase product longevity and to reduce reliance on natural resources (Lang & Armstrong, 2018; Elli et al., 2024). The fashion brands and retailers prefer to implement PSSs to enable the reuse of the materials and products. The specialized form of PSS for fashion retail is also known as the Fashion-as-a-Service (FaaS) model. Some researchers prefer to include the rental platforms in FaaS models. But the FaaS business model differentiates itself from the rental platforms in collaborative consumption models with the major objective of offering personalized fashion recommendations in addition to providing convenient and sustainable access to a wide range of garment options (Travasi & Musso, 2023).

2.2 *Sustainability in the Fashion Supply Chain*

The companies declare their commitments to sustainability in accord with the global targets and legislations. In this context, they employ sustainable practices in every stage of production and transform their supply and value chain into a more sustainable one. This section details the sustainability practices in preproduction, production, and postproduction stages, and the sustainability practices in supply chain management for transparency and traceability by giving examples.

2.2.1 Sustainability Practices in Preproduction

The fashion supply chain starts with the selection and sourcing of raw materials. The fashion suppliers prefer to use natural and biobased fibers due to their superiority in being biodegradable and preventing microfiber pollution (Tam & Lung, 2023). The traditional natural fibers such as linen, hemp, bamboo, kapok, wool, silk, alpaca, mohair, and angora are commonly preferred by fashion brands. Some of the natural fibers gained popularity recently because of their lower environmental impact. In this context, Levi's has presented the "Wellthread Collection" by using hemp fiber, which requires fewer chemicals and less water (Levi's, n.d.). Moreover, the Banana Republic has launched a sustainable fashion collection by using organic banana fiber (Hussain, 2023).

In addition to traditional natural fibers, the fiber companies develop novel fibers derived from natural or bio-based materials. For example, the company Temtex produces chitosan-based fibers, whereas Swicofil produces soybean protein fiber (TômTex, n.d.; Swicofil, n.d.). QMILK produces milk-based fibers, having the characteristics of being natural and as soft and smooth as silk (QMILK, 2018), and

Spiber produces plant-based material, Brewed Protein™ fiber made of sugar from sugarcane (*Spiber.inc.*, n.d.). Not only plant-based fibers but also bioplastics-based fibers are developed by fiber companies. In this regard, Algenesis is one company that has developed a fully biodegradable, high-performance polyurethane material by using Soleic® technology (Mayfield et al., 2020).

The fashion brands use these alternative natural fibers and novel bio-based materials in their collections. In this regard, Adidas' Futurecraft innovation platform has produced a biodegradable sneaker composed of biobased silk fiber by German AMSilk (Fernández, 2022). Likewise, the company Vollebak presented a degradable t-shirt manufactured using eucalyptus, beech pulp, and algae (Cairns, 2021). Fashion brands also opt for new bio-based materials as an alternative to leather material to protect animal rights. As is well known, Stella McCartney is a fashion brand that has committed to using artificial leather for handbags and shoes. The brand recently has reported that it uses BioCir® Flex, a nontoxic, compostable, recyclable, and biobased plastic that can be dyed with natural cinnamon (Stella McCartney, n.d.-a). Similarly, French luxury house Hermès has recreated its Victoria shopper bag by using an alternative vegan leather from mycelium, Sylvania, a biodegradable material with a low carbon footprint. Another mycelium-based alternative, Mylo, has been announced to be included in the products of Adidas, Stella McCartney, Lululemon (Hahn, 2021).

On the other hand, recycled fibers are used to reduce environmental impact and improve product sustainability (Islam et al., 2021). With respect to this, H&M launched several collections from recycled polyester, cotton, or Tencel. Specifically, it has introduced a collection from an artificial fiber, Circulose, derived from recycled cotton waste (Dragomir & Dumitru, 2022). Primark uses recycled cotton from textile waste by working in collaboration with Recover (Bi'Sektör, 2022). In the same vein, Inditex has made a three-year commitment to buy upcycled textile fiber Infinna™, which is generated from clothing waste (Inditex, 2022).

Even some fashion companies have initiated or taken a role in sustainability campaigns or cooperated with recycling companies to recycle the plastic waste. Inditex has given support to five recycling hubs in Europe in order to collect and recover textile waste (Dragomir & Dumitru, 2022). ECOALF initiated the project of "Upcycling the Ocean," which aimed to collect plastic waste in the oceans and remanufacture it for fashionable clothes and shoes (Cai & Choi, 2020). Kering Group has partnered with ECONYLVR, which produces nylon made from plastic waste (Tam & Lung, 2023).

Sustainable design is also an important preproduction activity. In this context, the companies discourage consumption and make timeless designs. For instance, Patagonia initiated the "Don't Buy This Jacket Campaign" to make customers reconsider buying new products unless absolutely necessary (Patov, 2024). Since 1993, Filippa K has had a mission to make consumers adopt buying fewer but better pieces to wear and love for years (*Filippa-K*, n.d.). Similarly, Eileen Fisher concentrates on simple and timeless designs, avoiding the take-make-waste manufacturing model (Eileen Fischer, n.d.).

It is also common that the companies give information about caring for garments for a longer lifetime. Being a fashion brand with timeless designs, Ralph Lauren gives information about the care of products in order to increase the durability of products (Ralph Lauren, n.d.). Nike explains specific caring instructions for Nike Forward (Nike, n.d.-a). Similarly, Nudie Jeans presents detailed information for the right care of products made up of 100% organic cotton (Nudie Jeans Co., 2020).

Other companies adopt the eco-design or zero-waste design principles. In this respect, Patagonia uses organic cotton, recycled fabrics, and traceable materials in its products (Cai & Choi, 2020). Levi's utilizes life cycle assessment to design products (Islam et al., 2021). Nike adopts circular solutions in design with Move to Zero by eliminating single-use plastics in products and converting textile waste to textile and non-textile products (Nike, n.d.-b). Similarly, H&M focuses on circularity in several collections, as in the case of its Cradle-to-Cradle certified NewBorn collection.

The minimization of garment parts or the use of accessories to ease recyclability are also adopted by some companies (Daukantienė, 2023). For instance, Organizzazione Vendite Speciali (OVS) reduces the number of garment components for maximum recyclability (Dragomir & Dumitru, 2022). Reformation refines the designs to consist of mono materials, making them easier to be recycled (Reformation, 2023). On the other hand, Weekday uses deconstructable accessories that can be reused in other garments (Weekday, 2023).

2.2.2 Sustainability Practices in Production

The textile manufacturing processes pose a serious pollution risk due to their intensive water, energy, and chemical usage. The majority of companies concentrate on the reduction of water consumption in wet processing like washing, dyeing, and finishing (Islam et al., 2021). In this context, the companies replace wet processing with dry processing practices and use closed-loop systems or automated dosing machines at different stages of textile finishing (Wren, 2022). In relation to this, Levi's has introduced a "waterless dyeing process" and a new laser technology to reduce water consumption and produce ripped and washed jeans more eco-efficiently (Cai & Choi, 2020). Jeanologia has reduced water consumption utilizing the closed-loop water system H2Zero (*Zero Waste Water Technology H2 Zero.*, n.d.). Not only the denim producers but all types of manufacturers opt for sustainable practices in wet processing and finishing. Nanushka has launched a new vegan alternative, which is produced by using a dry method that requires 80% less water compared to wet processing methods (Chan, 2022). VF Corporation and Inditex make their sourcing from the Zero Liquid Discharge suppliers (Dragomir & Dumitru, 2022).

The reduction of energy consumption and chemical discharge are the other priorities. Besides, less harmful substances are used in the chemical processes. A Turkish denim producer, Çalık Denim, uses Oxygene technology, which treats denim about 50–60% faster by using smaller quantities of chemicals and consuming less energy (Çalık Denim, n.d.). Tommy Hilfiger uses lower-impact finishes (Dragomir & Dumitru, 2022), and The North Face prefers non-perfluorinated chemicals (TheNorthFace, n.d.).

Alternatively, sustainable practices such as harvesting rainwater, using renewable energy sources, and consuming natural dyes are the other options preferred by the fashion brands. In this respect, a Turkish premium brand, Network, gives priority to manufacturers who develop rainwater harvesting systems in the production processes as a part of its N-Green project (Network, n.d.). Patagonia ensures the products were produced by using more renewable energy in accord with the bluesign® standard (Patov, 2024). A Turkish company, Natural Dye, offers natural dyes derived from plant waste materials such as walnut shells, acorns, orange peels, and pomegranate (Natural Dyes, n.d.), whereas the Spanish brand Atelier Aletheia creates unique colors from plant-derived dyes (Aletheia, 2025).

Sustainable textile and apparel production technologies play a pivotal role in the reduction of environmental damage by especially decreasing energy and material consumption. For instance, the replacement of old rapier looms with air jet weaving machines in denim weaving results in less energy consumption (Islam et al., 2021). The apparel wastes can be prevented by proactive practices with the use of CAD/CAM technology and automation (Islam et al., 2021). 3D software achieves zero-waste fashion design by digitalizing the design and eliminating the production of prototypes (Casciani et al., 2022). In this respect, The North Face teaches creation teams to design in 3D (Kleinman, 2022). Creating garments digitally reduces the carbon footprint by 97% and saves 3300 liters of water (Casciani et al., 2022).

The modern technologies, such as 3D printing and VR and AR technologies, further improve sustainability (Islam et al., 2021). 3D printing reduces the required time in design stages and minimizes waste in the production of 3D accessories (zips, buttons, trims, accessories, embellishments, etc.) and specific types of products (Tam & Lung, 2023). With respect to this, Adidas, Nike, and Under Armour integrated additive manufacturing into their production lines (Hohn & Durach, 2021). Balenciaga produced tailored coats and suit jackets for the Autumn/Winter Fashion Show 2018 and produced 3D-printed high heels with the customer's name on them in 2021 (Tam & Lung, 2023). AR and VR technologies are used to establish digital fashion shows, eliminating the need for transportation of products to introduce their collection for the new season ahead. In fact, the carbon footprint was reduced by 241,000 tons in 2020 with the restriction of the transportation of collections during international fashion weeks during the Covid-19 pandemic (Casciani et al., 2022). Some brands continue their activities on digital platforms. Concerning this, Christian Dior has launched a 360-degree view virtual store, and Balenciaga has introduced its Spring 2022 collection on a virtual runway (Tam & Lung, 2023). On the other hand, 3D and AR technologies are also planned to be employed for mass customization to reduce the overproduction, overconsumption, and stockholding of raw materials (Casciani et al., 2022).

Moreover, the integration of AI and data science-based technologies, such as IoT technologies, RFID, and robot systems in the apparel supply chain operation, prevents waste and overproduction and ensures the forecasting and production processes are efficient (Tam & Lung, 2023). Using RFID and integration of IoT solutions in production eliminates errors made by human factors, optimizes machine performance, and rationalizes energy use. AI can optimize the transportation and distribution, resulting in a reduction of energy consumption and air pollution with the real-time data. Moreover, the use of robots can speed up the production, enhancing the flexibility at the same time. Concerning these, Adidas uses the combinations of AI, RFID, IoT, and advanced analytics to optimize the inventory and the routes for distribution (DigitalDefynd, 2025). Besides, Adidas has two production facilities called "Speedfactories" in Germany and the US, which heavily rely on AI-driven robotics and automation (Hohn & Durach, 2021).

Being an indispensable part of production creating waste, packaging materials are also drawing special interest from fashion brands. In this respect, the use of hazardous chemicals in packaging is aimed to be reduced or replaced with eco-friendly materials (Islam et al., 2021). Specifically, the bio-based plastics, which can decompose under various environmental conditions, are used to decrease the environmental impact (Tam & Lung, 2023). Besides, recycled materials and Forest Stewardship Council (FSC) certified paper are used to decrease the virgin material requirement and support sustainable consumption. With respect to this issue, Gucci phased out virgin single-use plastics from packaging in 2020. Similarly, Kering prevented the use of polyvinyl chloride and glue in packaging, opting instead for recyclable packaging (Tam & Lung, 2023). The ready-to-wear brand, House of Sunny, has taken many measures. Specifically, it uses packages made from 100% compostable bio-based films, leaving no harmful chemical waste behind. In addition, the company uses logos and care labels made of recycled polyester, swing tags made of organic cotton, and price stickers from recycled cotton (House of Sunny, n.d.). In a similar vein, H&M has replaced single-use plastics with FSC-certified paper in packaging (Dragomir & Dumitru, 2022). Some retailers also found solutions for hanger materials. In this respect, Inditex has developed a single hanger used to transport garments from suppliers to stores, and Gap has invested in a hanger recycling system (Dragomir & Dumitru, 2022).

2.2.3 Sustainability Practices in Postproduction

Sustainability practices in postproduction cover the transportation and distribution operations and post-consumption processes. Minimizing transportation is a key priority in the postproduction stage, with efforts focused on reducing carbon emissions. In this respect, some companies prefer nearshoring and local shoring since the selection of suppliers closer to the company's physical location offers a reduced carbon footprint (Tam & Lung, 2023). For instance, an Istanbul-based mindful company, Hidden Bee, uses locally produced fabrics to ensure a lower carbon footprint (A Hidden Bee, n.d.). The cooperation with green suppliers in closer locations also increases the sustainable sourcing potential (Cai & Choi, 2020). Outerknown prioritizes vertically integrated or localized supply chains to reduce carbon emissions and prefers to work with the suppliers using renewable energy and water and employing chemical management programs (Outerknown, n.d.). The carbon emissions are also reduced with the determination of the optimum transportation option and the use of hybrid, electric, and biomethane-fueled vehicles (Tam & Lung, 2023; Islam et al., 2021). House of Sunny claims to always use sea freight over air freight (House of Sunny, n.d.), since shipping by sea results in carbon emissions corresponding to one thirty-fifth of those caused by air transportation (Niinimäki et al., 2020).

The other sustainability practices in the postproduction stage are about the reuse, resale, and recycling activities. These activities usually start with the implementation of take-back programs and reverse logistic activities, which are accomplished with the participation of end consumers (Paras & Pal, 2024). The collection of garments is usually made in stores. For instance, Inditex, Gap, and H&M have clothing collection points in stores. Even Gap and H&M encourage the return of used clothes by giving shopping credit (Dragomir & Dumitru, 2022). Some companies gather the product via post. Based on its Take-Back program, Patagonia receives products either physically or by post and sends anything made with cotton, hemp, or linen to the company, Infinited Fibers, to be turned into its recycled fibers to be remade into new T-shirts (Roberts, 2022). Not all collected garments are recycled; the ones that are in good condition are donated, increasing their lifetime. For instance, Uniqlo collects the post-consumer products for donation (Cai & Choi, 2020). Zara enables the reselling or donating of the products via the "Join Life" program (ZARA, n.d.).

Additionally, some companies either collaborate with secondhand and rental services or offer these services through their own platforms. Concerning this, Gucci and Alexander McQueen have collaborated with The RealReal and Vestiaire Collective, respectively (Tam & Lung, 2023). "Patagonia Worn Wear," "Levi's SecondHand," "Nike Refurbished," and "The North Face Renewed" are the official second-hand platforms owned by the companies themselves. H&M provides renting services in some stores, and Banana Republic rents customers a variety of garments with the "Style Passport" program (Wren, 2022). In the renting services, the laundry, pick-up, and delivery can also be organized through a third party (Paras & Pal, 2024). For instance, Nanushka has teamed up with the platforms Rent the Runway and Hurr to rent out its pieces. (Chan, 2022).

Finally, it is common for the brands to give repair services for consumers or present upcycled designs from used or unsold clothing. Specifically, Patagonia offers free or low-cost repairs with 62 repair centers worldwide (Patov, 2024). In Nudie Jeans Repair Shops, the jeans are fixed for free (Nudie Jeans Co., n.d.). Levi's reconstructs the vintage Levi's into modern fits and launches these designs in the Re/Done collection (RE/DONE, n.d.). Some brands have collaborated with third-party companies to facilitate the repair services. GANNI has partnered with SOJO to offer you a free tailoring and repairs service (GANNI, n.d.). PVH has a program called "Tommy for Life" in collaboration with The Renewal Workshop, in which the items are collected from consumers and either cleaned and repaired or upcycled and sold in the "Remixed collection" (Dragomir & Dumitru, 2022). Finally, Outerknown turns textile waste into wearable, one-of-a-kind upcycled pieces with Project Vermont (Outerknown, n.d.).

2.2.4 Sustainability Practices in Supply Chain Management for Transparency and Traceability

Putting genuine effort into the supply chain, discussed in the previous three sections, ensures that the supply chain becomes more sustainable. However, it is also important to ensure transparency and traceability for both business customers and end consumers. Therefore, achievements in both environmental and social sustainability aspects are certified through standards and certifications such as Global Organic Textile Standard (GOTS), Organic Content Standard (OCS), OEKO-TEX Standard 100, EU Ecolabel, Global Recycled Standard (GRS), Cradle to Cradle Certification, Responsible Wool Standard (RWS), Responsible Down Standard (RDS), Fair Trade International, Fairwear Foundation Certification, Certified B Corporation (B Corp), and Social Accountability (SA8000) to improve reliability. Moreover, companies support renewable energy usage and regenerative agricultural initiatives in order to achieve their carbon-neutral goals, reduce environmental impacts, and improve social welfare, such as the Better Cotton Initiative (BCI), the United States Department of Agriculture (USDA) Organic Certification, Cotton Made in Africa, and Business Social Initiative (BSCI).

Other tools used to improve transparency and traceability include the implementation of supply chain management software (such as SAP, Ariba, and Oracle SCM Cloud), the establishment of digital data-sharing platforms like Higg Index, and the use of sustainability tracing software and tools (Eco-chain, Sourcemap, Eco-Age, and B Corp). In this context, H&M, Nike, Adidas, and Levi's have employed Higg Product Tools to assess and measure the environmental and social performance of products and value chains through lifecycle assessments. VF Corporation uses traceability maps to reveal the product's origins and discuss end-of-life options for reduced impact (Dragomir & Dumitru, 2022). Moreover, several companies have invested in building their own tools and platforms. Inditex has a digital platform and an in-house Traceability Management System for suppliers that integrates all phases of the product's life cycle and provides information to stakeholders regarding the certification of fibers and water use per product. Pangaia has developed its own environmental management system to monitor energy and water consumption and waste generation (Pangaia, n.d.).

The final consumers are also allowed to reach the data in regard to sustainability via specific platforms and digital product passports. In this context, the digital platform "Footprint Chronicles" by Patagonia enables consumers to trace the environmental and social impact of each product (Patov, 2024). Digital Product Passports, providing information about composition, location, and manufacturer to supply chain stakeholders and consumers, have gained popularity in enhancing transparency and traceability recently (Paras & Pal, 2024). The methods used to link the DPPs to the physical product include Quick Response Codes (QR Codes)/Near Field Communication (NFC) Tags, Cloud-Based Data Sharing, and Blockchain. Partnering with a leading company in IoT technology, Alexander McQueen has designed the unique "My McQueen Concept," in which each garment is equipped with an NFC tag that is recorded digitally throughout the brand's supply chain until the consumer experience (Tam & Lung, 2023). Pangaia features QR-code-based digital passports so customers can see how sustainable their garments are (Ho, 2022). Nonetheless, blockchain applications take the lead over all other solutions in DPPs.

First proposed by Nakamoto in 2009, blockchain technology combines data records (encrypted blocks) within a network of actors (chains) by using data mining techniques (Oguntegbe et al., 2023). Blockchain can be employed from sourcing and production to distribution and consumption by providing advantages in terms of cost, speed, risk control, and flexibility (Chen, 2024; Wang et al., 2020). The historical performance data for suppliers helps to select suppliers and develop smart contracts (Wang et al., 2020). The performance records of potential supply chain partners can be tracked together with their environmental impact and footprints in a transparent platform after the smart contracts are activated. The working conditions with ethical criteria can also be recorded in a real-time manner. By integrating the ERP systems, a centralized order-production coordination can be provided by avoiding overproduction (Chen, 2024). The orders can be easily tracked, traced, and updated from upstream to downstream.

Besides, blockchain can be utilized for the reverse logistics and second-hand item processes. The entire reverse logistics process, such as collecting, reusing, and refurbishing waste, can be easily monitored by blockchain. The waste can also be reused by the integration of different circular supply chains. Specifically, two companies from different circular supply chains needing each other's waste can make transactions and free exchanges (Wang et al., 2020). The blockchain technology can be used to trace the second-hand products with all production and usage information. Providing disposal information, blockchain might facilitate recycling (Chen, 2024).

Based on these mentioned characteristics, blockchain improves sustainability and helps to manage a sustainable supply chain. However, its superiority in DPP applications goes beyond its reliability. The visibility, security, and transparency of blockchain promote trust in business operations (Wang et al., 2020). The events and achievements are digitally recorded in a chronological order and continuously updated by a network of actors. The smart contracts written in digital nodes eliminate the need for control by third parties, but it is not possible to influence the transaction without the consent of the other partners in the blockchain network (Oguntegbe et al., 2023). The encrypted and distributed ledger makes the blockchain fault resistant and immutable (Wang et al., 2020; Chen, 2024). Because of this reason, many companies are adopting blockchain-based DPP applications. In this respect, the Lenzing Group has launched Fibercoin to ensure data visibility and integrity throughout the entire ecosystem in partnership with Textile Genesis (Chen, 2024). Stella McCartney has partnered with UNECE on a blockchain technology to provide documentary proof of the production practices involved through certifications and standards (Stella McCartney, n.d.-b).

2.3 *Effect of Covid-19 Pandemic on Sustainable Fashion Production and Consumption*

The highly infectious Covid-19 virus was announced to cause a global pandemic by the World Health Organization on 11 March 2020 (McMaster et al., 2020). The Covid-19 pandemic changed the supply and demand dynamics by inducing short-term as well as long-term effects. The short-term effects have been particularly effective during the outbreak of the spreading of the pandemic. They emerged because of the specific containment measures, such as physical/social distancing and wearing face masks, recommended by the World Health Organization to prevent the widespread diffusion of the Covid-19 virus (Rinaldi & Bottani, 2023). The long-term effects have continued to be effective even after the pandemic's impact has diminished.

As short-term effects, the operations of businesses were stopped due to lockdowns, and this caused a temporary halt for the global economy for an unseen future in the earliest period of the pandemic. The uncertainty in the markets, sectors, and businesses influenced the global economy (Rinaldi & Bottani, 2023). The production facilities were closed or allowed to be open if they complied with certain restrictions, and the orders were cancelled and/or modified (Henninger et al., 2021). The import/export activities were limited as the borders were closed. In this respect, many obligations were imposed and measures were implemented, which forced companies and corporations to implement prompt actionable steps. In accord with these, many measures were taken in any type of business, such as those; the layout of offices and workspaces was reorganized with protective barriers between workstations, digital technologies were employed and meetings were established online, and the number of workers in the production shifts was reduced to avoid the simultaneous presence of numerous workers. On the other hand, a decrease in demand occurred because of restrictions, limitations of movements, and less frequent visits to stores (Rinaldi & Bottani, 2023). Besides, the fear of Covid-19 and feeling uncertain about the economic situation made consumers hesitate when they were making their purchases. Specifically, the demand for some product categories, particularly nonessential goods, decreased suddenly, whereas the demand for food and medical products showed a rapid rise.

Over time the short-term effects have become clarified, and it was observed that the severity level of outcomes varied for different sectors like tourism, trade, health, manufacturing, education, transportation, and fashion. The textile industry became one of the hardest-hit industries, along with the automotive and electronics industries, because of producing and selling nonessential goods and being export-oriented. The small and medium-sized enterprises were more affected by the disturbances because of their lower level of available resources and financial capability (Rinaldi & Bottani, 2023). Since the fashion supply chain has a fragmented structure and is widespread globally, the lockdowns and restrictions on borders resulted in delays in the production activities and the flow of products. Besides, the operational constraints in logistics led to higher freight rates (Tam & Lung, 2023).

The travel restrictions made it impossible to reach the available workforce and prevented the orders from being fulfilled (Rinaldi & Bottani, 2023). The pandemic also became a social crisis point as many workers lost their jobs and had financial struggles at that time (Henninger et al., 2021). For instance, the unemployment rate increased to 5.7% in February 2020 in China (Chakraborty & Biswas, 2020).

The problems in the supply chain also influenced the retailers in the end. Due to delays in production and distribution, the lead times were extended, and the stores became out of stock, causing a poor service performance (Rinaldi & Bottani, 2023). The misalignment in inventory forced the fashion companies to make extra discounts (Tam & Lung, 2023). Naturally, the consumption amount decreased. The clothing and footwear consumption reduced by more than 75% in the first quarter of 2020 (Tang, 2022). Clothing exports decreased 9.1% in 2020 compared to the previous year. The profits for global fashion companies have fallen by around 93% in 2020. Neiman Marcus Group, J.C. Penny, J. Crew, and Brooks Brothers have filed for bankruptcy protection; Nike, Gap, and H&M suspended their operations both online and offline (Tam & Lung, 2023); and the stock prices fell more than 50% for Macy's and Nordstrom and 25% for Adidas and Nike between February and March (Youn et al., 2021). Due to lockdowns and store closures, 376 Primark stores in 12 countries were shut down, accounting for a loss of some $751.5 million of net sales per month. On the other hand, the promotional activities were influenced by social distancing; for instance, the fashion events were no longer possible (Tam & Lung, 2023).

As a long-term effect, the Covid-19 pandemic changed the business landscape, highlighting the importance of adapting and reacting quickly and setting up mechanisms for crisis management in periods of uncertainty. The companies attempted to increase the resilience of their production and distribution systems (Rinaldi & Bottani, 2023). Specifically, the sourcing decisions were updated with the introduction of multiple or local sourcing strategies instead of single sourcing practices (Belhadi et al., 2022). The inventory levels were updated with the integration of safety stocks by a smaller amount of fashion companies (33%). Keeping safety stock was not adopted by all companies, however, as the fashion products capture the trends of the moment (Rinaldi & Bottani, 2023). On the other hand, the Covid-19 pandemic became a turning point in the acceptance of the deficiencies in terms of the digitalization of operations.

This understanding was not only initiated because of the challenges in the supply chain but also the demands of the final consumers. The consumers had to use the digital platforms to meet their needs. The online platforms almost became the only method for reaching consumers, and the pandemic underlined the importance of being online (Koch et al., 2020; Tran, 2021). Thus, the leading companies shifted to e-commerce for product distribution (Tam & Lung, 2023). The significance of the integration of digital tools into the supply chain to provide satisfactory service and on-time order fulfillment was also understood. Therefore, a positive inclination towards investing in data and information sharing occurred for enhancing the online presence and the interconnectedness in the supply chains (Rinaldi & Bottani, 2023).

Furthermore, the consumers got the consciousness about the environmental impact of their consumption and became more mindful of the environmental consequences (Henninger et al., 2021; Ong et al., 2021). Although they did not concentrate on fashion products but on sustainable consumption, Severo et al. (2021), Ali et al. (2021), and Valenzuela-Fernández et al. (2022) all confirmed the effect of the COVID-19 pandemic on environmental awareness and sustainable consumption in countries like Brazil, Portugal, Malaysia, and Latin America. Šimek and Sadílek (2024) showed that the period during the pandemic had a positive effect on the view of sustainability and the circular fashion economy by Generation Z consumers in Czechia. A notable shift occurred towards sustainable choices. Shen (2024) shared the findings of a longitudinal study over a period of 9 years in order to track the progress of consumer awareness of sustainable fashion. The results revealed that consumers were more aware of different product categories in sustainable fashion. Okur et al. (2023) assessed the effect of the Covid-19 pandemic on sustainable fashion consumption behavior by establishing a multigroup analysis and revealed that the consumers who were affected more by the Covid-19 pandemic had a greater concern and more environmental knowledge, which resulted in a higher tendency for sustainable fashion consumption. According to a consumer survey established in 15 global markets in April 2020, 54% of consumers stated they would make a permanent shift in making more sustainable choices (Brydges et al., 2020).

Specifically, fashion began to be viewed differently than before. The consumers increasingly express concerns about the sustainability of clothing products, such as the sources of raw materials and sustainable production and the welfare of workers during the pandemic. This consciousness was also provided via the social media platforms, which were highly visited by the consumers during the Covid-19 pandemic for reasons other than online shopping. In fact, consumers started to spend more time than ever on social media (Tam & Lung, 2023). The reputation of the fast fashion industry, prioritizing the economic performance and advertising expenditures, was criticized more with the increasing awareness of consumers about generating a remarkable amount of pollution (Wu et al., 2023a).

Investigating the perceptions and attitudes of Brazilian consumers regarding secondhand clothes comparatively before and during the Covid-19 pandemic, Amaral and Spers (2022) found out that Brazilian consumers were increasingly looking for more socially and environmentally responsible clothing items after the pandemic. In spite of the positive tendency towards purchasing sustainable fashion products, not all sustainable fashion products were equally embraced. The consumers avoided the swap shops, rental platforms, and second-hand stores, which are a part of circular business models, due to infection and hygiene concerns (Henninger et al., 2021; Brydges et al., 2020). Moreover, the second-hand stores for charity purposes were unable to collect and accept donations, leading the amounts of garments landfilled to increase.

Other than the limited number of exceptions, the Covid-19 pandemic is certain to have long-term effects in terms of increasing awareness for sustainability. The adoption of environmentally friendly lifestyles and digital markets by consumers assigned extra responsibility to fashion suppliers and emphasized the deficiencies of the fashion industry in regard to sustainability and digitalization, enabling the Covid-19 pandemic to be a transient event with enduring effects.

3 Sustainable Fashion Products and Consumer Behavior

Numerous studies that changed in terms of scope, sample characteristics, and the conceptual framework have been established about consumer behavior towards sustainable fashion consumption in literature. The Theory of Planned Behavior (TPB), the Theory of Reasoned Action (TRA), the Norm Activation Model (NAM), the Value-Belief-Norm (VBN), the Motivation Theory, and the Attachment Theory were the most common models used in these studies. The studies examined sustainable consumption with either an overall approach concentrating on sustainable fashion products in general or a specialized approach for different sustainable fashion product categories as described in Sects. 2.1.2 and 2.1.3. Therefore, this section includes four subsections dedicated to the consumer behavior analyses. The first three subsections include the studies about sustainable fashion product consumption in general, the products made up of environmentally friendly materials (including the ethical fashion products), and the slow fashion and collaborative consumption business models (such as secondhand and rental clothes), respectively. The last subsection includes the studies about consumer characteristics and sustainable fashion consumption. To show the consequences of the Covid-19 pandemic on sustainable fashion consumption, only the studies established between the years 2020 and 2025 are considered in each subsection, which ends up with the summary of suggestions established by the authors based on the findings of the studies.

3.1 Consumer Behavior Towards Sustainable Fashion Products

The studies aiming to understand the motivations beyond the behavioral intention of consumers towards the sustainable fashion products revealed that feelings, thoughts, and social norms were the most influential predictors of consumer behavior. A study established in Vietnam put forward that positive attitudes in terms of beliefs and feelings had the highest impact on the purchase intention towards the sustainable fashion products (Ho et al., 2020). Bianchi and Gonzalez (2021) claimed that the main driver of sustainable fashion adoption was the desire to contribute to a better world among Chilean female consumers. In the same vein, Nicolau et al. (2024) showed that feeling shameful was important to sustainable fashion purchase intention among Brazilian fashion consumers. Moreover, the consumption of sustainable products was considered a moral responsibility, and the opinions of other people were stated to be considered while making decisions. Davis and Dabas (2021) and Carfora et al. (2021) indicated that social norms influenced personal norms and in turn affected the purchase behavior. Banytė et al. (2023) revealed that social norms influenced the relationship between purchase intention and actual behavior in Lithuania. Using attachment theory as a base, Penz and Drewes (2022) showed that the opinions of peers enhanced the motivation to do something good for the environment. Similarly, Ou et al. (2023) and Floriano and Matos (2022) confirmed the significance of moral obligations, and Guan and Noor (2021) found out that customer endorsement together with ethical beliefs were the main predictors of sustainable fashion consumption among Millennial consumers in Malaysia.

In another group of studies, awareness, knowledge, and environmental concern were underlined to have a significant contribution to behavioral intention. Studying with a sample group of consumers aged between 18 and 50 in Indonesia, Nadyasworo et al. (2023) confirmed the positive correlation between awareness and the willingness to pay for sustainable fashion products. Similarly, Ersoy and Fu (2021) showed that the Canadian millennial consumers with a higher awareness were willing to pay more for sustainable fashion products, and Saeed et al. (2023) found out that the consumer willingness to purchase sustainable clothing among the participants aged between 16 and 35 in Pakistan was influenced positively by the impact of knowledge. Moreover, numerous researchers confirmed the notable impact of knowledge about environmental apparel and environmental issues on purchase intention (Jin & Omar, 2023; Ho et al., 2020; Ou et al., 2023; Han et al., 2024; Blazquez et al., 2020; Djossouvi et al., 2024). The environmental concern was also stated to be influential on behavioral intention. Having established two studies in Peru and Ecuador, Leclercq-Machado et al. (2022a, 2022b) found out that the purchase intention was influenced by the environmental concern and perceived environmental knowledge. Likewise, Sharma and Pal (2020) confirmed the strong correlation between environmental concern, environmental knowledge, and attitude, underlining the fact that consumers might have a higher concern for the environment even though they had a limited level of knowledge about environmental issues in the textile and apparel industry.

Apart from the studies investigating the motivations beyond the behavioral intention, the other studies examined the main drivers and barriers in the adoption of sustainable fashion products. Rahman and Koszewska (2020) implied that consumers made the purchase decision by considering not only the sustainable apparel cues, such as environmental and social/ethical aspects of apparel products, but also the non-sustainable apparel cues, such as functional, aesthetic, symbolic, and financial characteristics, in Poland. Similarly, Tryphena and Aram (2023) revealed that Indian consumers still considered the functionality, appearance, and price benefits. Investigating the barriers and motivations for sustainable fashion products in three continents, namely Europe, Asia, and North America, Brandão and da Costa (2021) revealed that the beliefs about poor style, quality, fit, and fashionable characteristics of sustainable fashion products and the availability of various styles and fast fashion items in the market still acted as major barriers for sustainable fashion consumption. In fact, Bocti et al. (2021) implied that the attitude-behavior gap mainly occurred because of a lack of fashionability, limited availability, and higher prices. Specifically, Rausch and Kopplin (2021) implied that sustainable clothing was still perceived as unfashionable and consumers cared about the aesthetic worries over the economic ones. Therefore, Jung et al. (2020) suggested offering attractive and authentic sustainable apparel products in China. Another study established among young Chinese females, Ma et al. (2024), underlined the limited number of alternative styles and distribution channels for sustainable fashion products and added that higher price levels also discouraged consumers from purchasing sustainable fashion products. Stephan and Chugh (2021) confirmed the expensiveness of sustainable fashion products. In parallel with this, Tryphena and Aram (2023) reported that the sustainable fashion products were expensive but expected to be priced lower if the material was locally sourced.

Even though it was not mentioned as a main barrier or driver, the impact of greenwashing claims and trust for sustainability were also considered in many studies in recent years, highlighting the fact that consumers were suspicious about the environmental claims and they desired the reliability of the companies. To be more precise, Jin and Omar (2023) and Rausch and Kopplin (2021) showed that the greenwashing concern had a negative impact on both attitudes and purchase intentions. Having employed the extended TPB model with consumer effectiveness in order to understand the intentions of Turkish consumers to purchase sustainable fashion, Shehneh et al. (2024) pointed out that the inconsistencies and gaps, such as greenwashing and misinformation, discouraged the consumers from purchasing sustainable fashion products. On the contrary, Hong et al. (2024) confirmed that trust was a significant predictor of sustainable fashion consumption behavior among Chinese adults by using the value-belief-norm framework.

Based on the findings of the studies regarding the positive impact of attitudes and social norms on behavioral intention, the authors advised establishing campaigns involving messages about the positive contribution of sustainable fashion consumption to enhance sustainable fashion consumption in general. Moreover, the majority of authors suggested establishing social media campaigns with the participation of influencers and celebrities, in which consumers are encouraged to share their own stories/experiences on the social media platforms. The studies confirming the significance of awareness, knowledge, and concern on sustainable fashion consumption indicated that the awareness and knowledge level of end-users should be enhanced by the help of government and policymakers. In this context, it has been advised that policymakers develop more targeted policy measures to promote the achievements of sustainable consumption and incorporate sustainability issues into the school curriculum. Regarding the findings of studies concentrated on the drivers and barriers for sustainable fashion consumption, the researchers advised the fashion brands to present products that foster consumers perception of style, fashion, quality, and fit. The high price perception of sustainable fashion products was suggested to be altered either by applying some discounts in the form of incentives or by shifting the mindset of consumers from buying low-priced, low-quality items towards buying fewer quality items with increased life cycles. Finally, some suggestions were made for the companies and governments about the greenwashing concerns and trust issues. In this respect, the companies were advised to provide transparency in releasing the information, like suppliers, manufacturing processes, and business certificates. Specifically, it was recommended that the clothing retailers publish an annual sustainability report certified by independent auditors and take action to implement green initiatives. Nonetheless, the governments are stated to play a leading role in demanding transparency from the fashion industry and in imposing strict penalties when false information is disclosed.

3.2 Consumer Behavior Towards Environmentally Friendly Materials and Ethical Fashion Products

As in the case of sustainable fashion products, the impact of beliefs and attitudes, particularly social norms and altruism, on behavioral intention was emphasized for the environmentally friendly products. Abrar et al. (2021) revealed that the most influential factor in purchasing eco-friendly products was attitude, followed by subjective norms for Generations Y and Z in Pakistan. Likewise, Islam and Sultana (2020) showed the positive influence of moral obligation on young Bangladeshi consumers' purchase intention towards biodegradable clothing. Rinaldi et al. (2024) found out that the purchase intention for eco-friendly clothing by Generation Z was significantly influenced by altruism in Indonesia. Similar findings were obtained by Farooqi and Frooghi (2024) in terms of green apparel in Pakistan. Moreover, Shin and Lim (2021) highlighted that the effect of altruism on behavioral intention towards eco-friendly fashion products and Tewari et al. (2022) confirmed the attitude was influenced positively by altruism for green apparel, and finally, Farzin et al. (2023) revealed that social influence had the major impact on purchase intention of eco-fashion products, followed by altruism, respectively.

In parallel with the findings of studies about sustainable fashion products, concern and knowledge were found to be important in consumer behavior towards environmentally friendly materials and processes. Mathur (2024) revealed that the purchase intention of Generation Z was significantly influenced by environmental concerns. Employing the NAM and SOR algorithms in combination, Tandon et al. (2023) found out that the green apparel buying intention of consumers in the USA was influenced positively by ascription of responsibility and ecological concern. Moreover, Sobuj et al. (2021) and Elmanadily and El-Deeb (2022) indicated that the eco-friendly apparel purchase intention was influenced by both environmental concern and environmental knowledge. Kumar et al. (2021) stated that consumers will be willing to pay more if the environmental consciousness is enhanced with the explanation of benefits in an appropriate way. Liu et al. (2024) confirmed the positive impact of environmental awareness and social responsibility on the green apparel intention of Generation Z consumers. Similar findings were established by Lim and Lady (2023) for Generations Y and Z in Batam.

Other researchers put greater emphasis on the influence of environmental knowledge. Zhang et al. (2024) investigated the influential mechanism underlying the purchase of green clothing in China and revealed that consumers' level of environmental knowledge had a significant positive impact on purchase intention. Rahman and Koszewska (2020) claimed that eco-clothing would not be widely accepted when consumers lacked the "green" knowledge. Khare et al. (2020) showed that the knowledge about organic clothing had a positive effect on both organic clothing product involvement and consumption involvement, and Khare and Kautish (2022) showed that green apparel knowledge and green apparel perceived effectiveness had a positive influence on Indian consumers' green apparel purchase behavior. Moreover, Khare (2023) revealed that the perceived benefits of green apparel were influenced by green apparel knowledge. Sadiq et al. (2021) found out that the optimistic consumers bought the green apparel, while the environmental knowledge could decrease the pessimism. The knowledge was also handled from the perspective of being related to health benefits. Venkatraman et al. (2021) showed that consumers' awareness of health and hygiene had a positive impact on the purchase of sustainable antimicrobial textiles made of organic fibers with herbal finish. Moreover, organic apparel knowledge was found to alter the perceived quality and health benefits perception about organic apparel (Davis & Dabas, 2021).

As in the case of sustainable fashion products, the price and aesthetic properties of the environmentally friendly products were shown to negatively influence the behavioral intentions towards environmentally friendly apparel. Specifically, Tandon et al. (2023) revealed that American consumers considered sustainable products expensive and indicated that the buying intention towards green apparel was negatively influenced by green apparel affordability. Kumar and Mohan (2021) expressed price sensitivity as a significant negative predictor of environmentally friendly apparel purchase intention. A study established among Korean consumers by Shin and Lim (2021) showed that the consumer confusion because of price sensitivity and greenwashing had a negative effect on purchase intention towards eco-friendly fashion products. Moreover, an interview-based study established by Munir (2020) identified the limited availability and the unattractive styles of eco-fashion clothing as the most important barriers. Additionally, Neupane et al. (2021) stated that the aesthetic claims should be explicit, whereas the environment-friendliness claims should be implicit since aesthetic concerns were more important than environmental concerns.

Unlike previous studies established for sustainable fashion products, the perceived values were examined in terms of their contribution to the consumption of environmentally friendly products. In this context, Pandey and Yadav (2023) validated the effect of perceived value on consumption attitude toward green apparel. Liang et al. (2022) verified the positive impact of altruistic, biospheric, and egoistic values on the green apparel consumption behavior of Chinese Generation Z consumers. Other researchers concentrated on the utilitarian, environmental, and hedonic values. In this respect, Bielawska and Grebosz-Krawczyk (2021) showed that the Polish consumers' behavior was positively influenced by emotional, conditional, and environmental values. Rūtelionė and Bhutto (2024) confirmed the utilitarian and environmental values changed attitudes toward green apparel. Muraguri et al. (2020) verified the positive influence of hedonic value, utilitarian value, and environmental value on consumers' perceived green value of green home apparel in China. In parallel with these, Durrani et al. (2023) and Lin et al. (2023) showed the impact of hedonic and utilitarian motivations on the behavioral intention towards eco-friendly apparel products.

There are also some studies in literature that associate the consumption of environmentally friendly products with status value, self-concepts, and brand engagement. Specifically, Zhu (2021) showed that the Chinese millennial consumers preferred eco-fashion products because of their status value. Abrar et al. (2021) revealed that green self-concept determined the purchase intention of Generations Y and Z. Anisah et al. (2024) verified the positive and significant influence of green self-concept on the purchase intention of environmentally friendly clothing. On the other hand, Lee and Lee (2024) showed that the brand-self-connection changed the behavioral intentions, indicating that brands are perceived as tools for revealing oneself.

Moreover, eco-labeling, social media, and influencers were shown as triggering factors in environmentally friendly product consumption. Dhir et al. (2021) showed that the satisfaction with labeling positively influenced the purchase intentions. Khare et al. (2022) drew attention to the impact of celebrities on green clothing purchase behavior. Similarly, Kautish and Khare (2022) showed that social media influence, peer influence, cosmopolitanism, and global social identity were influential on green apparel knowledge, which in turn determined sustainable fashion apparel perception of Indian consumers.

The consumption of ethical fashion products was investigated in a limited number of studies. Blazquez et al. (2020) revealed that Spanish consumers fostered higher positive intention for ethical fashion than environmentally friendly fashion. On the other hand, Vlastelica et al. (2023) investigated the determinants of environmentally and socially responsible clothing consumption of young adult customers (18–35 years) in Serbia and found that green consumption values were the most important determinant of responsible apparel consumption. Liu et al. (2021) used the TPB model and showed that the subjective norm and the attitude toward ethical fashion products predicted purchase intention, whereas the knowledge of ethical fashion played a substantial role in trust toward ethical fashion products. Bae and Yan (2023) focused on values and observed that the ethical values failed to instill a socially responsible mindset in individuals that were heavily immersed in fashion or materialism.

Based on the extensive discussion about the importance of knowledge and values in both environmentally friendly fashion and ethical fashion, the authors have reached the conclusion that the standards and certifications are very important for enhancing consumption of these products. Therefore, it was advised to increase the awareness of consumers about these standards with some type of educational campaign or explanations from sales personnel. Besides, it was strongly suggested to put a special emphasis on health-related benefits of organic apparel and natural dyes, such as less skin irritation and the reduction of skin contact with harmful chemicals. Since it was put forward that some consumers associate environmentally friendly consumption with their self-concept, brand engagement, and opinion leaders, it was recommended that brand image should be shaped with the commitment to sustainability. Furthermore, green clothing consumption should be projected as the latest fashion trend via online communities, blogs, microblogs, podcasts, and wikis. It was stated that consumers could be reassured about the quality of materials used in sustainable apparel products, and therefore they are worth the price. Finally, it was advised that the final products be presented with attractive designs and a wide variety incorporating new fashion trends, fabrics, and styles.

3.3 Consumer Behavior Towards Slow Fashion Products and the Collaborative Consumption Business Models Such as Second-Hand and Rental Clothes

Few studies were established about slow fashion products between the years in concern. The beliefs, attitudes, or perceived value became the examined factors in these studies. Specifically, Chi et al. (2021) showed that the attitude and perceived consumer effectiveness significantly affected US consumers' intentions to purchase slow fashion apparel. The authors added that US consumers would purchase slow fashion apparel if they believed that they could contribute to the protection of the environment. In parallel with this, Moodly et al. (2023) indicated that slow fashion was adopted by consumers who are aware of the environmental damages and social consequences of fast fashion. Moreover, Lira and Costa (2022) revealed that slow fashion consumption was influenced by ethical considerations in Brazil. Seock et al. (2024) found out that knowledge about slow fashion had a high impact on the environmental sustainability attitudes of Generation Y and Z in the USA market. Similarly, Li-Hsun and Jerusalem (2021) implied that the awareness about the environmental issues was influential on the psychological attitudes towards slow fashion. On the other hand, Castro-López et al. (2021) indicated that the consumer orientation toward slow fashion consumption depended on the positive perceived value of slow fashion.

The fashion and exclusivity characteristics of slow fashion also drew a special interest from researchers. Specifically, Suhud et al. (2020) revealed that slow fashion orientation triggered fashion involvement, which in turn resulted in purchase intention and willingness to pay a premium price. Underlining the positive impact of environmental, utilitarian, and hedonic values on purchase intention, Aprianingsih et al. (2023) stated that the consumers could be convinced by the everlasting design characteristics of slow fashion products. Accordingly, Siaw et al. (2024) stated that the exclusivity, authenticity, functionality, localism, and equity characteristics of slow fashion were important for the consumer preferences.

On the other hand, there is still indecision about slow fashion adoption. Araújo et al. (2022) stated that slow fashion was in a proto-practice state among millennial women, whereas Faludi (2022) indicated that ethical, slow, eco-conscious, second-hand, and vintage may all constitute a growing market.

The number of studies regarding the collaborative consumption-based business models is higher compared to the ones regarding slow fashion products. Since these models are considered new because of transforming consumers to users and shifting the acquisition experiences to using experiences (Henninger et al., 2021), the studies about the collaborative consumption-based business models investigate the motivations lying beyond consumers approaches, the potential of these models to be adopted by consumers, and, finally, the drivers and barriers lying beyond the adoption of them.

As in the case of other types of sustainable products, the studies investigated the impact of beliefs and attitudes related to environmental considerations. In this respect, Fabiana et al. (2024) and Borusiak et al. (2020) revealed that the purchasing of secondhand products was influenced by the awareness of environmental problems and sense of responsibility. Accordingly, Chaturvedi et al. (2020) showed that environmental concern and personal norms were influential on the purchase intention of recycled clothing positively. Nonetheless, not only the positive attitudes but also the negative emotions, such as feeling guilty, were addressed as the motivations lying beyond the purchase intention towards secondhand clothing by some researchers (Wu et al., 2023b; Syahrivar et al., 2023).

Other researchers drew attention to the importance of social and subjective norms. Hoang et al. (2022) indicated that the subjective norm was the remarkable determinant of purchasing second-hand clothing. Similarly, Vicamara et al. (2023) revealed that social support had a positive impact on shopping intention towards used clothing in the Indonesian market, and Hasbullah et al. (2022) stated that the subjective norm was positively linked to the purchase intention of upcycled fashion products. Moreover, Koay et al. (2022) showed the importance of injunctive norms, descriptive norms, and moral norms on consumers' intention to buy second-hand clothing in Malaysia. Finally, the attitude and purchase intention towards upcycled and recycled clothing were stated to be influenced by the core self-oriented, social self-oriented, and object self-oriented motivations in the given order of magnitude (Dwikesumasari et al., 2024).

The perceived value of collaborative consumption-based models was also investigated in some studies. Specifically, Yang et al. (2024) confirmed the positive impact of environmental values on ecological worldview, which would result in preference for these models. The other studies drew attention to the utilitarian and hedonic values. In this respect, Tymoshchuk et al. (2024) revealed that the purchase intentions and word-of-mouth behaviors of US consumers were positively influenced by the perceived hedonic, utilitarian, and environmental value perception of second-hand clothing in the given order of magnitude. In accord with this, Kim et al. (2021) stated that hedonic benefit and trust in the product history positively influenced the attitude and use intention. Specifying the hedonic values, Wang et al. (2022) expressed that the young people born in the 1990s and 2000s in China were significantly driven by treasure-hunting fun.

There were also some studies investigating the impact of motivations other than the ones mentioned above. In this regard, Lang and Zhang (2024) showed that the intention towards fashion renting was associated with frugality and perceived enjoyment. Sharma et al. (2024) indicated that the motivations for purchasing second-hand clothing had ideological, economic, and convenient bases.

The rest of the studies attempted to determine the future potential of the collaborative consumption-based models, as well as the main drivers and barriers in the adoption of them. de Aguiar Hugo et al. (2023) stated that these novel models have the potential to be adopted by especially younger consumers due to their willingness to change their shopping habits and beliefs that the fashion brands should be more sustainable. Focusing specifically on rental clothing in the UK, Herold and Prokop (2023) implied that older generations would also adopt such practices, just more slowly. However, Musova et al. (2021) stated that the awareness level of the consumers about these models was still low based on the study established among Slovak consumers.

The majority of the studies pointed out the price and quality characteristics as the main drivers of adopting secondhand products. In this respect, Taylor et al. (2023) identified key drivers of second-hand clothing purchasing for young women in the Australian market as the cost savings in addition to their environmental factors. Focusing on the consumer-to-consumer online platforms for second-hand clothing, Yeap et al. (2024) found out that the economic motivations, situational frugality, and sustainability motivations positively affected the attitude towards second-hand fashion products. Specifically, Zahroh (2023) indicated that the price-saving orientation was the most important motivator for purchasing second-hand clothing. Similarly, Abutaleb et al. (2023) revealed that the attitudes towards the sharing economy were influenced by the economic benefits. Additionally, Sajjad et al. (2021) implied that consumers preferred buying second-hand clothes since they were cheaper and usually good quality, and Stancu et al. (2020) implied that the sustainable clothes were positioned as long-lasting and unique by consumers. Koay et al. (2022) indicated that the low-income consumers preferred secondhand clothing because of economic reasons, but high-income consumers preferred secondhand clothing because of their needs for exclusivity. On the contrary, the quality issues and the limited availability of alternative sizes of the unique products were listed as drawbacks by Taylor et al. (2023).

The characteristics of secondhand products being fashionable were highlighted in some studies. In this respect, fashion leadership and the need for uniqueness were shown to be the main drivers of using apparel rental services among Generation Z in the USA (McCoy et al., 2021). The positive impact of fashion motivation and originality on purchase intention towards secondhand fashion products was also confirmed by Aycock et al. (2023) and Tangri and Yu (2023), respectively. Moreover, Herjanto et al. (2023) revealed that the fashion innovativeness had triggered the attitudes toward secondhand clothing in addition to the materialism and financial pressure. Kala and Chaubey (2024) defined novelty-seeking, fashion orientation, and narcissism as the main drivers for behavioral intention toward fashion clothing rental in India. Having investigated the effects of Chinese young consumers' traits on behavioral intention towards using clothing subscription rental services, Su et al. (2023) listed the three main factors as fashion leadership, price consciousness, and social media usage.

The other studies revealed the adoption barriers of secondhand and rental clothing were basically social risks. In this context, Debnath (2023) attempted to understand how Generation Z consumers in India perceived the use of secondhand clothing and found out that the social context was the primary driving factor, but the fear of penalization in the social setting was a major hindrance in the adoption of secondhand clothing. Similarly, Silva et al. (2021) indicated that social embarrassment had the highest negative influence on the purchase of secondhand clothing. Moreover, Sajjad et al. (2021) explored social influence as one of the main reasons for avoiding secondhand clothes, and Zahroh (2023) indicated that social norms negatively moderated the relationship between attitude and purchase intention. Furthermore, Lang and Zhang (2024) expressed that the perceived social risk had a negative impact on attitude and perceived enjoyment. Azzahra and Purwanegara (2024) indicated that the purchase intention towards secondhand products was negatively influenced by the perceived risk for Indonesian consumers too. Similarly, Kala and Chaubey (2024) implied that the perceived risk had a negative contribution to consumers' attitudes towards rental clothing. According to Wang et al. (2022), the concern about poverty was among the main purchase barriers for secondhand clothing. Finally, Yeap et al. (2024) also confirmed that the performance risk and social risk negatively moderated the relationship between attitude and intention to purchase second-hand clothing.

The other barrier was associated with the hygiene issues, especially during the Covid-19 pandemic. Sajjad et al. (2021) indicated that hygiene was found to be one of the main reasons to avoid buying second-hand clothes. Wang et al. (2022) and Xu et al. (2022) showed that the sanitation conditions of secondhand clothing prevented the adoption of secondhand clothing. On the other hand, Silva et al. (2021), who investigated the impact of product characteristics in purchasing secondhand clothing for experienced and inexperienced consumers, revealed that the experienced consumers could overcome some traditional barriers, such as peers' opinions, the store environment, animosity, and hygiene.

Based on the findings of the studies about slow fashion consumption, the clothing designers and manufacturers were advised to concentrate on the timeless design and longevity characteristics of slow fashion products and to provide premium design and quality with high artistic value. Moreover, it was suggested to develop collections including easy-to-mix-and-match items, enabling the existing clothes to be used for a longer time period with distinctive and personalized styles.

The findings about the collaborative consumption-based models put forward that they were not preferred because of the environmental reasons but also because of economical, utilitarian, and hedonic purposes. Nonetheless, they struggled with the negative perception of the consumers in regard to social and hygienic risks. Considering these findings, it was recommended by the authors to enhance the image of secondhand and rental clothing to cover sustainability. It was specifically advised that the platforms use embedded messages about saving the environment and inform, remind, or reinforce the enlightened consumers' belief that they have made the right choice by buying second-hand clothing. Utilization of social media, collaborations with celebrities, bloggers, and influencers, and establishment of storytelling campaigns were strongly recommended to enhance consumer awareness and engagement. The utilitarian benefits were suggested to be satisfied by keeping a robust inventory, giving a convenient service, and presenting more information about products, such as the year/season of the collection, the chief designer of the time, and the origin of the country, by creating a narrative for a product history. The hedonic benefits were suggested to be improved by creating a playful retail store layout and displaying the clothes in racks or on mannequins rather than in piles, enhancing the appearance. Besides, the store managers and salespersons were suggested to give some idea about upcycling this secondhand and rental clothing. In regard to price considerations, the companies were advised to incorporate different price points for consumers in distinctive financial brackets and offer discounts or sales promotion incentives to motivate repeat purchases. On the other hand, it was also advised that the platforms sell only secondhand clothing that is clean and in good condition.

The rental companies were suggested to emphasize the claims of novelty, uniqueness, exclusivity, and high-status fashion clothing to attract novelty-seekers and young consumers, who can express their fashion orientation through aspirational brands.

Some suggestions were also made to governments. The governments were advised to consider offering tax relief to shoppers who bought second-hand apparel or to arrange campaigns to educate consumers about the circular economy and collaborative consumption and encourage them to play an active role in the sustainable disposal of excess clothing. The schools were stated to take a lead role in teaching basic repair skills and proper care in the right way in reselling clothing through various channels.

3.4 Consumer Characteristics and Sustainable Fashion Consumption

There are some studies in literature investigating the impact of sociodemographic characteristics on consumer behavior. One of the consumer characteristics examined is the gender. In this context, Chi et al. (2021) revealed that the female consumers were found to show a more positive attitude toward slow fashion apparel. Similarly, the female consumers who belonged to Generation Z had a greater inclination towards sustainable fashion consumption in comparison to the male consumers (Zhang et al., 2023). Moreover, Dangelico et al. (2022) stated that women had a higher purchase intention and higher willingness to pay for the apparel products made up of organic materials. On the other hand, the interviews established by Bianchi and Gonzalez (2021) with Chilean female consumers revealed that the eco-conscious women preferred to buy sustainable fashion products from secondhand retail stores since sustainable fashion brands were considered high-end fashion products with high prices.

In regard to age, younger people were given a greater focus in the consumer behavior studies. In this respect, Dangelico et al. (2022) revealed that younger consumers would have a higher willingness to pay for sustainable clothing. Besides,

Musova et al. (2021) stated that the younger consumers had a higher willingness to support slow fashion. Hur and Faragher-Siddall (2022) added that young consumers strongly support the eco-labeling and educational policies. On the other hand, Gurova (2024) revealed that sustainability was considered fashionable among young consumers and the secondhand products were preferred because of lower prices. Kalia and Kaur (2021) found out that the majority of respondents, 58.1%, did not wish to purchase used clothing, and only a few, 16.2%, said they occasionally wanted to buy or borrow apparel, however.

Generations were given a special consideration in the consumer behavior studies. Based on a survey conducted in India, Gupta et al. (2023) confirmed that the millennials were concerned about the harmful effect of fast fashion on nature and society. Similarly, Rasmussen (2020) revealed that millennials possessed an affinity for sustainability and had the feelings of "irritation" and "guilt" when purchasing fast fashion. Making comparisons between different generations, Pandey and Yadav (2023) confirmed that Generations Y and Z were slightly different in terms of green apparel involvement. Lin and Chen (2022) stated that Generation Z showed a stronger association with rental apparel, second-hand apparel, and recycled apparel in comparison to Generations X and Y in Taiwan.

Since one of the main characteristics of Generation Z consumers was stated to be their sensitivity about sustainability, Generation Z consumers were given a special emphasis in the studies. In this respect, Kovacs (2021) examined the consumer behavior of Generation Z from different perspectives. In this respect, Generation Z consumers were stated to associate sustainable clothing with environmentally friendly production, cruelty-free production, local production, the less and reuse of fabric, products made up of organic cotton, and the support of domestic brands in the given order. Moreover, they were found to check the product labels and tags while making a purchase decision. Nonetheless, they were found to have a positive attitude towards the use of recycled materials in production, and avoid secondhand products depending on the negative emotions about the purchasing experience and used clothing. Similarly, redesigning, reusing, and upcycling old apparel products were not favored by Generation Z. Besides, the generation Z consumers were stated not to be sensitive about the waste reduction. Moreover, Palomo-Domínguez et al. (2023) delved into the motivations of Generation Z consumers towards buying and selling second-hand apparel products in digital sharing economy platforms in Lithuania. They revealed that they preferred secondhand products in order to meet environmental, social, and ethical needs, as well as the need for trendiness. Finally, Zhang et al. (2023) revealed that the passionate, eco-conscious Generation Z consumer actively searched for knowledge about sustainability and supported secondhand clothing stores.

Apart from the gender and age, the education and income were also investigated in a limited number of studies. Even though most of the studies did not find the moderating effect of these parameters, some studies revealed a discriminating influence. In this respect, Muraguri et al. (2020) and Durrani et al. (2023) stated that the income levels had a regulating influence on consumers' behavior. Chi et al. (2021) indicated that consumers with a higher income level were more likely to purchase slow fashion apparel. Interestingly, Hoang et al. (2022) stated that the higher-income individuals tended to buy more secondhand clothes. In regard to education, Wu et al. (2023b) implied that the impact of the perceived values of secondhand clothing varied for the consumer group of undergraduates and the group composed of participants having education below a master's degree. Dangelico et al. (2022) revealed that the consumers with higher education had a higher tendency to purchase sustainable clothing. On the other hand, they were found to be willing to pay more for only the organic products.

Finally, a limited number of cross-cultural studies were established about sustainable product consumption. In this respect, Han et al. (2024) revealed that the association of perceived value with the purchase intention was different for Korean and Chinese consumers. Working with Brazilian and Dutch consumers, Gomes et al. (2022) showed that the materialistic values did not explain circular behavior in the apparel industry among Brazilian respondents, but they had a positive impact on circular apparel behavior among Dutch consumers. Cross-cultural research established in the UK, US, and China by Jung et al. (2021) revealed that the utilitarian and hedonic values, pro-environmentalism, and perceived consumer effectiveness were more important for the US consumers than for Chinese and Korean consumers. Besides, UK and US consumers were more sensitive about the sustainability of brands, whereas the Chinese consumers were more sensitive about functional characteristics of sustainable fashion products. Brandão and da Costa (2021) conducted a survey among 669 consumers from Europe, Asia, and North America. The authors revealed that the perceived value was important in Asia and North America and the aesthetic characteristics were important in European markets. Duh et al. (2024) examined the adoption of the slow fashion idea in individualistic countries like Canada and South Africa and a collectivist country like China and showed that the information quality and trust were more problematic in individualistic cultures.

The studies investigating the sociodemographic characteristics of consumer behavior revealed that consumers of the same age and gender had similar tendencies. However, consumers across diverse cultural and geographical contexts exhibited distinctive tendencies in terms of beliefs and attitudes. Therefore, it is advised that the retailers consider these differences while they are building strategies for different markets.

4 Empirical Study

The literature review established in recent years showed that the approaches of consumers towards sustainable fashion products were positive and usually associated with the awareness of consumers about the environmental and social aspects of their consumption and the sense of responsibility towards themselves and others. Besides, the behavioral intentions might vary according to the sociodemographic characteristics. As mentioned in Sect. 2.3, the Covid-19 pandemic changed the shopping habits and behavior for both the short and long term. Considering the points above, the scientific literature might benefit from employing a model focusing on the impact of environmental awareness and social responsibility on behavioral intention, which had been triggered by the external factor, specifically the Covid-19 pandemic. Moreover, analyzing the impact of the pandemic on the behavioral intentions of different generations, which has not been considered in the previous studies, enhances the knowledge. Based on these, the purpose of this empirical study is to explain the sustainable fashion consumption of different generations, which has been expected to be triggered because of the Covid-19 pandemic. In this respect, the TPB model and the NAM model have been used in combination. Being an intensively used model to explain a person's pro-environmental behavior, the TPB was preferred since it assumes that most people's behavior is rational and the buying behavior is linked strongly to consumer beliefs (Borusiak et al., 2020; Sinha et al., 2022; Koay et al., 2022). The NAM framework was preferred since it successfully examines the pro-environmental behavior derived from moral obligation or sense of responsibility (Shin et al., 2018). TPB addresses self-interest motives and explains someone's personal expectancy and benefits, while NAM addresses pro-social motives and explains the behavior derived from moral obligation or sense of responsibility (Abutaleb et al., 2023; Joo et al., 2022; Li et al., 2020). The combined model serves to explain the behavior in both self-interest and pro-social aspects, increasing the comprehensiveness. The integrated model was extended with an additional construct, the fear of the Covid-19 pandemic, to show if the Covid-19 pandemic triggered sustainable fashion consumption. Finally, the findings were evaluated for the two consumer groups, Generation Y and Generation Z, to see if their beliefs and level of being affected by the external events differed.

4.1 Research Model and Data Analysis

Providing air quality improvement as evidence, the Covid-19 pandemic became the starting point of broader sustainable awareness since it allowed the consumer to understand their role in environmental pollution and producing waste (Valenzuela-Fernández et al., 2022; Joo et al., 2022). This realization was confirmed with the studies established in countries Brazil, Portugal, and Latin America by Valenzuela-Fernández et al. (2022) and Severo et al. (2021). Specifically, the fear of the pandemic and the uncertainty because of the pandemic were stated to trigger the pro-environmental behaviors and take action for the protection of the environment. (Gao et al., 2020; Jian et al., 2020; Ng et al., 2020; Chwialkowska et al., 2020; Hofstede, 2001). Moreover, the Covid-19 pandemic was reported to affect society, as people come together in times of crisis, become sensitive to other people, and support each other with a higher social responsibility and generosity (Valenzuela-Fernández et al., 2022; Severo et al., 2021; Ali et al., 2021). The environmental awareness was claimed to have a positive attitude toward eco-friendly activities (Kim, 2023; Carfora et al., 2021; Floriano & Matos, 2022) and social norms (Sonnenberg et al., 2022; Wang et al., 2021). The feeling of responsibility was also stated to be an antecedent for social and/or personal norms according to the NAM theory (Woosnam et al., 2023; Kim, 2023; Tandon et al., 2023; Borusiak et al., 2020; Joo et al., 2022). Besides, the ascription of social responsibility was accepted to have a positive influence on reuse intention of second-hand clothing since it gives people a goal of knowing and doing what needs to be done (Wu et al., 2023b). Finally, there are many studies in literature that confirmed that the purchase intention was determined by both the attitude and subjective norms. The greater positive attitude might mean more engagement in sustainable fashion purchase intention (Sobuj et al., 2021; Blazquez et al., 2020). Based on these facts and assumptions, a conceptual model was built as seen, in Fig. 1, for which the hypotheses were given as below.

H1: Fear of the Covid-19 pandemic (FOC) has a positive impact on the environmental awareness of consequences (EAW).
H2: Fear of the Covid-19 pandemic (FOC) has a positive impact on social responsibility (SR).
H3: Environmental awareness of consequences (EAW) has a positive impact on subjective norms (SN).
H4: Environmental awareness of consequences (EAW) has a positive impact on attitude (ATT).
H5: Social responsibility (SR) has a positive impact on subjective norms (SN).
H6: Social responsibility (SR) has a positive impact on attitude (ATT).
H7: Subjective norms (SN) have a positive impact on purchase intention (PI).
H8: Attitude (ATT) has a positive impact on purchase intention (PI).

A questionnaire involving two parts was developed for the survey. While the first part included the questions about the sociodemographic characteristics of the participants, such as age, gender, education, income level, and purchasing habits, the second part included the questions about the approach of consumers towards the Covid-19 pandemic and sustainable fashion consumption. Eight questions were adapted to evaluate the fear of the Covid-19 pandemic (Jian et al., 2020; Tran, 2021; Severo et al., 2021). Six questions for environmental awareness (Severo et al., 2021), five questions for social responsibility (Severo et al., 2021), five questions for behavioral attitude (Han et al., 2010; Hansen, 2008), three questions for

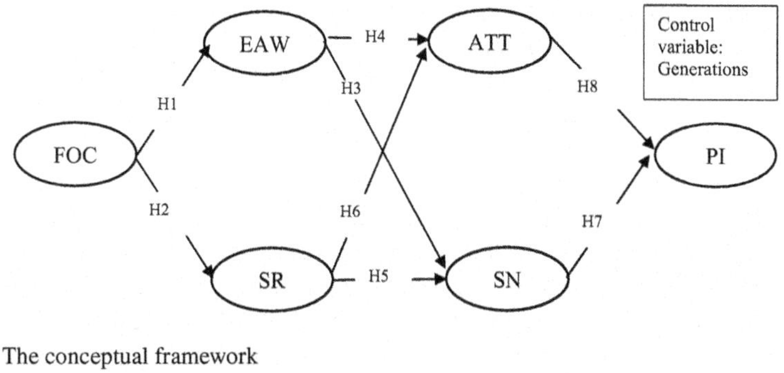

Fig. 1 The conceptual framework

subjective norms (Koch et al., 2020; Maichum et al., 2016), and five questions for purchase intention for green products (Maichum et al., 2016) were adapted from the studies in literature. The questions in the second part were established using a seven-point Likert scale. The data was collected via an online survey conducted with a target of reaching 500 participants, including participants belonging to Generations Y and Z in Turkey. IBM SPSS 27 and IBM AMOS 22 statistical programs were used to analyze the data sets. The univariate and multivariate normality were examined with the skewness and kurtosis values and Mardia's test, respectively. The factors were extracted by using exploratory factor analysis (EFA) and then confirmed by using confirmatory factor analysis (CFA). Some modifications were made by examining the modification indices in error terms of direct variables to improve the model fit. The internal consistency was examined by calculating Cronbach's alpha (CA) of each factor. The composite reliability (CR) and the average variance extracted (AVE) were calculated to check the convergent validity of each factor. A multigroup invariance test was performed to assess the equivalence in the measurement model between the groups, and the responses of the groups could be compared. The structural equation modeling (SEM) was used to test the hypotheses and to calculate the strength of the relationships between factors. The squared multiple correlations (SMC) of the dependent variables were calculated to examine the predictive accuracy of the model.

4.2 *Results and Discussion*

The data was collected from 507 participants. The number of male and female participants was almost equal to each other. 193 participants (38.1%) belonged to Generation Y, and 314 participants (61.9%) belonged to Generation Z. The majority of the participants were undergraduates (82.2%) for Generation Z and graduates (57.5%) for Generation Y. Both Generation Y (51.3%) and Generation Z (52.2%) members belonged to the medium income class, and more than half of the participants (59.1% of Generation Y and 56.4% of Generation Z) stated that they spend more than 10% of their disposable income on apparel expenditure. The demographic profile of Generation Y and Generation Z was given in Table 1.

The data set was checked for univariate normality, and no significant evidence for threats to normality was found. The scale reliability was satisfied since CA was found to be 0.94, which was well above the acceptable value of 0.6. The sample was found to be adequate, as KMO got the value of 0.927 and Bartlett's test of sphericity was significant (10503.608, DF.465, 0.001). The EFA was applied by using the principal axis factoring with varimax rotation. One item with factor loadings below 0.4 was deleted. The 31 items yielded six factors having factor loadings between 0.471 and 0.856, explaining 67.7% of the total variance.

The multivariate normality was assumed for the data set. The multivariate kurtosis got the value of 459.391 for 31 items, which was lower than the calculated value of 1023 (31*33). The CFA was established on the maximum likelihood estimation,

Table 1 Demographic profile of the sample

Categories	Groups	Gen Y	%	Gen Z	%	TOTAL
Gender	Male	94	48,7	166	52,9	260
	Female	99	51,3	148	57,4	247
	Total,%	193	38,1	314	61,9	507
Education	Under graduate	82	42,5	258	82,2	340
	Graduate	111	57,5	56	17,8	167
	Total,%	193	38,1	314	61,9	507
Income	Low	27	14,0	89	28,3	116
	Medium	99	51,3	164	52,2	263
	High	67	34,7	61	19,4	128
	Total,%	193	38,1	314	61,9	507
% of spending on apparel	Below 10%	79	40,9	137	43,6	216
	Above 10%	114	59,1	177	56,4	291
	Total,%	193	38,1	314	61,9	507

Table 2 The goodness-of-fit indices

The goodness of fit indices	Values used for comparison	Initial CFA	Final CFA	Final SEM	Group SEM
Chi-Square		1636.688	1015.216	796.901	1587.42
Degree of freedom		449	403	392	784
Chi-square/degree of freedom	≤3 (Kline, 2005)	3.645	2.519	2.033	2.025
Goodness of fit index (GFI)	≥0.9 (Hair et al., 1998)	0.822	0.884	0.907	0.839
Adjusted goodness of fit index (AGFI)	≥0.8 (Chau, 1996)	0.790	0.857	0.883	0.796
Comparative fit index (CFI)	≥0.9 (Hair et al., 1998)	0.888	0.940	0.961	0.927
Incremental fit index (IFI)	≥0.9 (Hu & Bentler, 1999)	0.888	0.941	0.961	0.928
Normed fit index (NFI)	≥0.9 (Hair et al., 1998)	0.852	0.905	0.926	0.867
Root mean square of approximation (RMSEA)	≤0.08 (Browne & Cudeck, 1993)	0.072	0.055	0.045	0.045
Root mean square residual (RMR)	≤0.08 (Browne & Cudeck, 1993)	0.071	0.053	0.059	0.074

whose results were given in the third column in Table 2. The model fit values were improved, as seen in the fourth column of Table 2, after checking the modification indices for error terms and correlating the error terms having high shared variance in order to improve the model fit.

CA, CR, and AVE values were calculated. CA values were between 0.84 and 0.92 for the measurement model, revealing an internal consistency as all the values are above 0.6. The convergent validity was confirmed as the CR and AVE values of all factors have higher values than the benchmarks of 0.6 and 0.5, respectively.

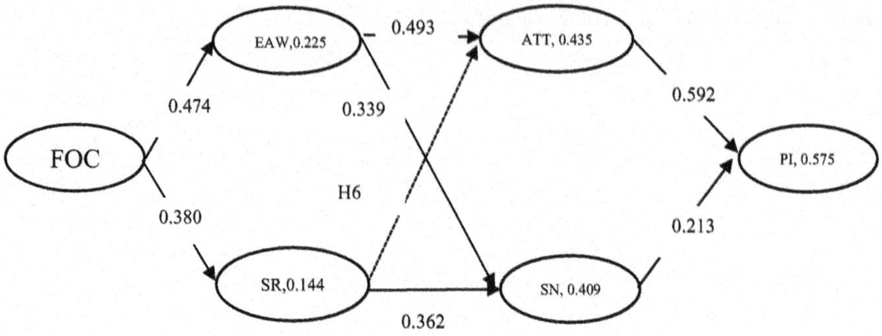

Fig. 2 The standardized regression weight for all participants

Table 3 The goodness-of-fit values of baseline and constrained model and results of invariance test

	CMIN	df	CMIN/df	CFI	NFI	RMSEA	Δdf	ΔCMIN	p-value
Baseline model	1589.718	782	2.033	0.926	0.867	0.045			
Constrained model	1633.781	807	2.025	0.925	0.863	0.045	25	43.963	0.011

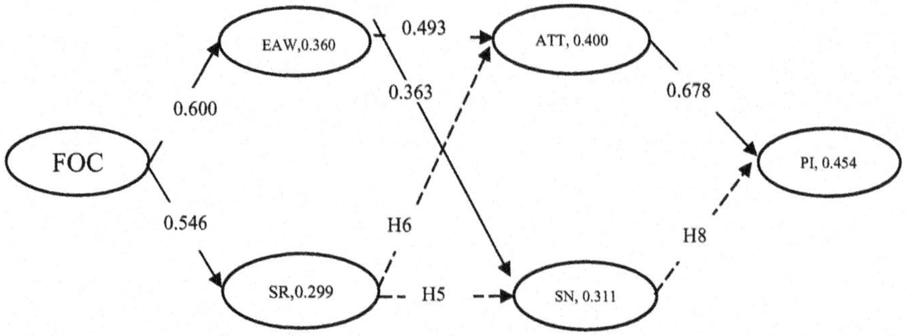

Fig. 3 Confirmed hypotheses, standardized regression weights, and R^2 values (* 0.05 significance level) for Generation Y

The SEM analysis was established for all participants whose results were given in the fifth column in Table 2. The standardized regression weights were given in Fig. 2. It was observed that all the hypotheses were satisfied except H6.

For the multigroup analysis, the metric invariance test was conducted, which resulted in a significant p-value (0.011), as seen in Table 3. This indicated that the groups were different at the model level.

The hypotheses were tested by using SEM. The conceptual model had the goodness-of-fit values, as shown in the sixth column in Table 3. The standardized estimates were generated for the relationship, and the SMC values were calculated for the dependent variables and demonstrated in Figs. 3 and 4 for Generation Y and Generation Z, respectively. For the Generation Y participants, all of the hypotheses except H5, H6, and H8 were confirmed at the 0.05 level of significance. FOC was

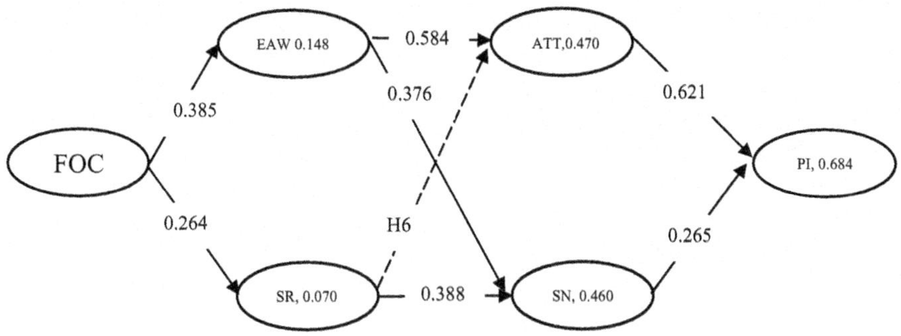

Fig. 4 Confirmed hypotheses, standardized regression weights, and R^2 values (*0.05 significance level) for Generation Z

found to be influential on both EAW and SR. EAW, but not the SR, was found to be influential on both SN and ATT, which in turn were influential on PI. For Generation Z participants, all of the hypotheses except H6 were confirmed at the 0.05 level of significance. In parallel with Generation Y participants, FOC was found to be influential on both EAW and SR, whereas EAW was found to be influential on ATT and SN, and SR was found to be influential on SN. Finally, both ATT and SN were found to have an impact on PI.

In the current study, the findings justified that FOC had a significant impact on both EAW and SR. Besides, the impact of FOC was higher on EAW than on SR for both generations. Reinforcing the impact of the Covid-19 pandemic on people's sustainable behaviors, Ali et al. (2021), Severo et al. (2021), and Valenzuela-Fernández et al. (2022) reached similar conclusions, stating that the Covid-19 pandemic had a significant and positive impact on the constructs of EAW and SR. In a similar vein, all the authors found out that the intensity of the impact of the Covid-19 pandemic on EAW was relatively higher compared with its intensity on SR. This result provided evidence for the fact that the Covid-19 pandemic improved that EAW and concerns about climate change, and it became a turning point for the transition of sustainable consumption and the reduction in air pollution (Ali et al., 2021). The less intensive impact of FOC on SR highlighted that the respondents were not engaged or concerned with social issues to that extent (Severo et al., 2021). On the other hand, when the intensity of the impact of FOC on EAW and SR was compared for Generation Y and Generation Z, it was observed that the values of both relationships were higher for Generation Y than for Generation Z in the current study. Since Generation Y is older than Generation Z, this result might be attributed to the fact that the older people perceived the Covid-19 situation as more severe, as they thought that they were more vulnerable to its consequences than younger people (Youn et al., 2021).

Findings of the current study indicate a significant positive impact of EAW on SN and ATT for both Generation Y and Generation Z in this study. Moreover, the strength of the relationship between EAW and ATT was bigger than the relationship between EAW and SN. The positive effect of EAW on SN was in parallel with the previous findings (Arkorful et al., 2021; Sonnenberg et al., 2022), putting forward the fact that the awareness of consequences led people to adopt beliefs of acting according to the social norms. The positive impact of EAW on ATT aligns with findings of preceding literature (Rastegari et al., 2023; Woosnam et al., 2023; Joo et al., 2022). Specifically, the influence of EAW on ATT was confirmed for sustainable fashion products among Brazilian consumers (Floriano & Matos, 2022) and for Italian female consumers (Carfora et al., 2021). In the studies in which both the relationship between EAW and ATT and the one between EAW and SN were investigated, the relationship between EAW and ATT was found to be higher compared to the one between EAW and SN (Arkorful et al., 2021), highlighting the fact that EAW causes changes in actions more than in beliefs.

SR was found to be influential on SN for only Generation Z, whereas it was not influential on ATT for both generations in the current study. The positive effect of SR on personal norms was carried out for sustainable clothing and pro-environmental behaviors by Tandon et al. (2023), Joo et al. (2022), and Kim (2023). Even though Joo et al. (2022) stated that this relationship was significant irrespective of age, the findings of this study put forward that consumers' feelings of responsibility would cultivate an obligation because of social connection to engage in sustainable apparel consumption for only Generation Z consumers. In regard to the findings of this study about the relationship between SR and ATT, they were in parallel with the findings of Arkorful et al. (2021) and Tandon et al. (2023). Nonetheless, there was a contradictory finding by Liang et al. (2022) for Generation Z, who revealed that Generation Z has the realization about their influence on environmental protection by finding a positive significant impact of SR on green consumption intention. The difference in Generation Z behavior might be attributed to the perspectives of different cultures.

The positive impact of ATT on INT was confirmed for both Generation Y and Generation Z in the current study, whereas the significant relationship between SN and INT was revealed for only Generation Z. The positive effect of ATT on INT was associated with the fact that consumers hold strong beliefs that sustainable apparel purchases are a good idea and extremely necessary. Without making a further analysis for generations, a significant relationship was also validated between SN, ATT, and INT (Sobuj et al., 2021; Leclercq-Machado et al., 2022a, 2022b; Kumar et al., 2021; Ong et al., 2021). Among these studies, the strength of the relationship between ATT and INT was found to be higher than that of the one between SN and INT (Leclercq-Machado et al., 2022a, 2022b; Kumar et al., 2021). Even Abrar et al. (2021), Kumar and Mohan (2021), Chi et al. (2021), Floriano and Matos (2022), and Rausch and Kopplin (2021) found out that the relationship between ATT and INT is significant, but the relationship between SN and INT is insignificant. Therefore, the findings of the current study are in accord with the findings of the previous studies. Specifically, the findings of this study provided evidence that the younger generation was more likely to behave in a more environmentally responsible manner, as expressed by Echegaray and Hansstein (2020).

4.3 Practical Implications and Recommendations

The current study theoretically implied a successful integration of the NAM and TPB models in the context of sustainable purchase intention and became the first study to investigate FOC as an external variable for the integrated framework. It was shown that sustainable fashion consumption was determined not only by ATT from the TPB model but also by SR and EAW from the NAM model, highlighting the contribution of combining elements of different theories for a deeper understanding of sustainable purchase intention. Finally, this study has deepened the comprehensive framework by investigating and confirming the moderating role of generation.

The results of the present study showed that the following:

- Fear of Covid has triggered both EAW and SR for both generations.
- EAW is the primary factor to be influential on both SN and ATT.
- ATT is a primary factor that helps the formation of PI towards sustainable apparel products. But SN is also influential on the purchase intention for Generation Z consumers.
- Generation Z also feels more responsible to the other people they are living with.

Overall, the importance of the main determinants, EAW and ATT, was common for the two generations in concern, but they were found to behave differently in terms of the factors related to altruistic behavior, SR and SN. This reconfirmed the fact that the marketing strategies adopted for promoting sustainable fashion products to Generation Y might not be effective for Generation Z consumers. Therefore, distinguishing strategies should be applied for different generations.

This study provided evidence that both Generation Y and Generation Z consumers in Türkiye have EAW about the ongoing issues of environmental degradation. This awareness can be improved by the marketers by employing advertisements that include rational and logical explanations about environmental degradation and the role of sustainable fashion product consumption on the betterment of the environment. Specifically, a more rational appeal in the promotional messages would be convenient for Generation Y, as altruistic behavior is not common for them, according to the findings.

The causal path between SR, SN, and PI for Generation Z suggests that Generation Z consumers have stronger ethical and moral values toward environmental protection. Therefore, the marketers and/or the advocacy groups of the companies should highlight the role of personal obligations for sustainable fashion consumption when they are developing the promotional messages for Generation Z consumers. The message content designed should provide information about social proof or stimulate a desire for social approval. The persuasive messages that instill a sense of responsibility among consumers and how their engagement with sustainable fashion consumption benefits others would be appropriate for Generation Z consumers.

Moreover, as SN was found to be influential on PI, Generation Z consumers might be encouraged to share their sustainable fashion consumption experience and information to promote sustainable fashion consumption and to build a positive consumption community within their circle of friends. As social media is an important source of information for Generation Z, the platforms or the channels, such as websites, blogs, and social network sites, might be chosen for delivering a campaign. For instance, the brands might invest in social media influencers to advertise on YouTube, TikTok, or Instagram to promote sustainable fashion consumption to attract customers of Generation Z.

Findings of this study also bear relevance for policymakers and provide some references for the promotion of sustainable fashion consumption around the world. The impact of FOC on both EAW and SR showed that the environmental values might be activated in a particular situation, especially in the context of a severe environmental crisis. This showed that environmental issues are not stable over the course of time and they can change with external events. Nonetheless, considering the significance of EAW on behavioral intentions, multilevel educational policies can be developed through the education system covering school education, corporate education, and family education. Education at different levels should be provided within a distinct concept (covering mostly the facts and realities or highlighting the individual responsibilities), as the values, sensitivities, and tendencies of generations differed from each other. Moreover, local governments might support companies offering sustainable fashion products by relaxing taxes or by providing easy loan facilities, awards, recognition, and regulatory flexibilities.

Theoretical implications of the current study lie beyond the effectiveness and the comprehensiveness of the integrated model in explaining behaviors. The proposed model extended with an external variable became the first model to deepen the framework in the domain of sustainable fashion consumption. The practical implications presented in this study might be useful for the authorities to provide a more responsible and friendly society toward the environment and guide the development and execution of strategies and actions to better communicate with the consumers.

However, this study is not free of limitations. But these limitations might provide avenues for future research. The first limitation of the study is that the study has been performed on a convenience sample of consumers, but future research would benefit when conducted with a wider and more appropriate sample by establishment of probabilistic sampling. The other point is that the study was established in one country, and so some results might be specific to this country. Having a well-based textile fashion and being a growing market with a young population, the establishment of this study in Türkiye might provide a comprehensive understanding of sustainable fashion consumption in any country with a similar context. However, it is advised to replicate the study in other countries to enhance the generalizability of the findings and to make a comparison between different countries, as the cross-cultural studies can make major contributions to literature.

Our study highlighted that the Covid-19 pandemic has made an impact on both EAW and SR, and future dynamics can bring other crisis situations, which might have similar contributions. The research can be established in a similar context since longitudinal studies can benefit from seeing the effect of changing dynamics. Future studies may consider the actual purchase behavior for sustainable fashion products, which might be equally insightful. Moreover, other moderator variables such as gender, education, or income might be added to explain some unexpected results.

Acknowledgement The authors would like to thank to Abdulmelik TURAN, Emir NALINCIOGLU and Mesut KARACA for their contribution in conducting the survey of empirical study as a part of their Senior Design Project.

References

A Hidden Bee. (n.d.). *About.* https://ahiddenbee.com/pages/about

Abbate, S., Centobelli, P., Cerchione, R., Nadeem, S. P., & Riccio, E. (2024). Sustainability trends and gaps in the textile, apparel and fashion industries. *Environment, Development and Sustainability, 26*(2), 2837–2864.

Abutaleb, S., El-Bassiouny, N., & Hamed, S. (2023). Towards a consumer social behavior theory for sharing economy: An integrated view. *Management & Sustainability: An Arab Review, 2*(1), 25–46.

Abrar, M., Sibtain, M. M., & Shabbir, R. (2021). Understanding purchase intention towards eco-friendly clothing for generation Y & Z. *Cogent Business & Management, 8*(1), 1–19.

Aletheia. (2025, January 22). *Atelier - Aletheia.* https://atelier-aletheia.com/en/atelier/

Alexa, L., Apetrei, A., & Sapena, J. (2021). The COVID-19 lockdown effect on the intention to purchase sustainable brands. *Sustainability, 13*(6), 3241.

Alhaddi, H. (2015). Triple bottom line and sustainability: A literature review. *Business and Management studies, 1*(2), 6–10.

Ali, Q., Parveen, S., Yaacob, H., Zaini, Z., & Sarbini, N. A. (2021). COVID-19 and dynamics of environmental awareness, sustainable consumption and social responsibility in Malaysia. *Environmental Science and Pollution Research, 28*(40), 56199–56218.

Amaral, J. H. G., & Spers, E. E. (2022). Brazilian consumer perceptions towards second-hand clothes regarding Covid-19. *Cleaner and Responsible Consumption, 5*, 1–14.

Anisah, T. N., Della Nanda Luthfiana, V. K., Ernestivita, G., Harnaji, B., & Najmudin, M. (2024). Green threads: Unveiling sustainable fashion preferences among Indonesian gen Z consumers. *Asian Management and Business Review, 4*(2), 376–393.

Aprianingsih, A., Fachira, I., Setiawan, M., Debby, T., Desiana, N., & Lathifan, S. A. N. (2023). Slow fashion purchase intention drivers: An Indonesian study. *Journal of Fashion Marketing and Management: An International Journal, 27*(4), 632–647.

Araújo, M. K. F. D., Mesquita, R. F. D., Matos, F. R. N., & Sobreira, M. D. C. (2022). Fashion consumption practices of millennials women: Between fast and slow fashion. *Revista de Administração da UFSM, 15*(4), 615–633.

Arkorful, V. E., Lugu, B. K., & Shuliang, Z. (2021). Unearthing mask waste separation behavior in COVID-19 pandemic period: An empirical evidence from Ghana using an integrated theory of planned behavior and norm activation model. *Current Psychology*, 1–16.

Arrigo, E. (2020). Global sourcing in fast fashion retailers: Sourcing locations and sustainability considerations. *Sustainability, 12*(2), 508.

Aycock, M., Cho, E., & Kim, K. (2023). "I like to buy pre-owned luxury fashion products": Understanding online second-hand luxury fashion shopping motivations and perceived value of young adult consumers. *Journal of Global Fashion Marketing, 14*(3), 327–349.

Azzahra, M. S., & Purwanegara, M. S. (2024). The effect of subjective norm and perceived risk on consumers' purchase intention towards secondhand clothes. *International Journal of Current Science Research and Review, 7*(8), 6373–6377.

Bae, S. Y., & Yan, R. N. (2023). Promoting the right attitude: Exploring the ethical fashion attitudes of fashion aficionados and materialists. *Journal of Fashion Marketing and Management: An International Journal, 27*(6), 1064–1087.

Balwan, W. K., Singh, A., & Kour, S. (2022). 5R's of zero waste management to save our green planet: A narrative review. *European Journal of Biotechnology and Bioscience, 10*(1), 7–11.

Banytė, J., Vaidelinskaitė, Š., & Šalčiuvienė, L. (2023). Investigating the link between consumer attitudes and behaviour in the context of sustainable clothing: The role of social norms. *Sustainability, 15*(24), 16800.

Bhandari, N., Garza-Reyes, J. A., Rocha-Lona, L., Kumar, A., Naz, F., & Joshi, R. (2022). Barriers to sustainable sourcing in the apparel and fashion luxury industry. *Sustainable Production and Consumption, 31*, 220–235.

Belhadi, A., Kamble, S. S., Venkatesh, M., Jabbour, C. J. C., & Benkhati, I. (2022). Building supply chain resilience and efficiency through additive manufacturing: An ambidextrous perspective on the dynamic capability view. *International Journal of Production Economics, 249*, 108516.

Bianchi, C., & Gonzalez, M. (2021). Exploring sustainable fashion consumption among eco-conscious women in Chile. *The International Review of Retail, Distribution and Consumer Research, 31*(4), 375–392.

Bielawska, K., & Grebosz-Krawczyk, M. (2021). Consumers' choice behaviour toward green clothing. *European Research Studies Journal, 24*(2), 238–256.

Bi'Sektör. (2022, August 7). Primark sürdürülebilir spor giyim serisini piyasaya sürüyor | Bisektör. *Bisektör*. https://www.bisektor.com/primark-surdurulebilir-spor-giyim-serisi/

Blazquez, M., Henninger, C. E., Alexander, B., & Franquesa, C. (2020). Consumers' knowledge and intentions towards sustainability: A Spanish fashion perspective. *Fashion Practice, 12*(1), 34–54.

Bocti, M., El Zein, S. A., & Giannini, R. (2021). Exploring antecedents to the attitude-behavior gap for sustainable fashion consumption in Germany. *Journal of Sustainable Marketing, 2*(2), 27–38.

Borusiak, B., Szymkowiak, A., Horska, E., Raszka, N., & Żelichowska, E. (2020). Towards building sustainable consumption: A study of second-hand buying intentions. *Sustainability, 12*(3), 875.

Botsman, R., & Rogers, R. (2010). *What's mine is yours: The rise of collaborative consumption*. Harper Business.

Brandão, A., & da Costa, A. G. (2021). Extending the theory of planned behaviour to understand the effects of barriers towards sustainable fashion consumption. *European Business Review, 33*(5), 742–774.

Browne, M. W., & Cudeck, R. (1993). Alternative ways of assessing model fit In: Bollen KA, Long JS, editors. Testing structural equation models. Beverly Hills. CA: Sage, 111, 135.

Brydges, T., Retamal, M., & Hanlon, M. (2020). Will COVID-19 support the transition to a more sustainable fashion industry? *Sustainability: Science, Practice and Policy, 16*(1), 298–308.

Cai, Y. J., & Choi, T. M. (2020). A united nations' sustainable development goals perspective for sustainable textile and apparel supply chain management. *Transportation Research Part E: Logistics and Transportation Review, 141*, 102010.

Cairns, R. (2021, October 26). *This vegan 'tech-style' startup wants to make clothes using algae*. CNNstyle. https://edition.cnn.com/style/article/israel-algae-environment-textiles-spc-intl-hnk/index.html

Carfora, V., Buscicchio, G., & Catellani, P. (2021). Integrating personal and pro-environmental motives to explain Italian women's purchase of sustainable clothing. *Sustainability, 13*(19), 10841.

Casciani, D., Chkanikova, O., & Pal, R. (2022). Exploring the nature of digital transformation in the fashion industry: Opportunities for supply chains, business models, and sustainability-oriented innovations. *Sustainability: Science, Practice and Policy, 18*(1), 773–795.

Castro-López, A., Iglesias, V., & Puente, J. (2021). Slow fashion trends: Are consumers willing to change their shopping behavior to become more sustainable? *Sustainability, 13*(24), 13858.

Centobelli, P., Abbate, S., Nadeem, S. P., & Garza-Reyes, J. A. (2022). Slowing the fast fashion industry: An all-round perspective. *Current Opinion in Green and Sustainable Chemistry, 38*, 100684.
Chakraborty, S., & Biswas, M. C. (2020). Impact of COVID-19 on the textile, apparel and fashion manufacturing industry supply chain: Case study on a ready-made garment manufacturing industry. *Journal of Supply Chain Management, Logistics and Procurement, 3*(2), 181–199.
Chan, E. (2022, July 22). *How Nanushka is making its vegan leather more sustainable*. British Vogue. https://www.vogue.co.uk/fashion/article/nanushka-new-vegan-leather
Chaturvedi, P., Kulshreshtha, K., & Tripathi, V. (2020). Investigating the determinants of behavioral intentions of generation Z for recycled clothing: An evidence from a developing economy. *Young Consumers, 21*(4), 403–417.
Chau, P. Y. (1996). An empirical assessment of a modified technology acceptance model. *Journal of management information systems, 13*(2), 185–204.
Chen, Y. (2024). How blockchain adoption affects supply chain sustainability in the fashion industry: A systematic review and case studies. *International Transactions in Operational Research, 31*(6), 3592–3620.
Chi, T., Gerard, J., Yu, Y., & Wang, Y. (2021). A study of US consumers' intention to purchase slow fashion apparel: Understanding the key determinants. *International Journal of Fashion Design, Technology and Education, 14*(1), 101–112.
Choi, D., & Hwang, T. (2015). The impact of green supply chain management practices on firm performance: The role of collaborative capability. *Operations Management Research, 8*, 69–83.
Chwialkowska, A., Bhatti, W. A., & Glowik, M. (2020). The influence of cultural values on pro-environmental behavior. *Journal of Cleaner Production, 268*, 122305.
Çalık Denim. (n.d.). *World needs more OXYGENE - CALIK DENIM*. CALIK DENIM. https://www.calikdenim.com/resources/events/world-needs-more-oxygene
Dangelico, R. M., Alvino, L., & Fraccascia, L. (2022). Investigating the antecedents of consumer behavioral intention for sustainable fashion products: Evidence from a large survey of Italian consumers. *Technological Forecasting and Social Change, 185*, 122010.
Daukantienė, V. (2023). Analysis of the sustainability aspects of fashion: A literature review. *Textile Research Journal, 93*(3–4), 991–1002.
Davis, L., & Dabas, C. (2021). Capturing sustainable fashion purchase behavior of Hispanic consumers in the US. *Journal of Global Fashion Marketing, 12*(3), 245–259.
de Aguiar Hugo, A., de Nadae, J., & da Silva Lima, R. (2023). Consumer perceptions and actions related to circular fashion items: Perspectives of young Brazilians on circular economy. *Waste Management & Research, 41*(2), 350–367.
Debnath, S. (2023). Perception of second-hand fashion in Indian context: Purchase decisions of generation Z. *Empirical Economics Letters, 22*(2), 71–89.
Dhir, A., Talwar, S., Sadiq, M., Sakashita, M., & Kaur, P. (2021). Green apparel buying behaviour: A stimulus–organism–behaviour–consequence (SOBC) perspective on sustainability-oriented consumption in Japan. *Business Strategy and the Environment, 30*(8), 3589–3605.
DigitalDefynd, T. (2025, March 11). *5 Ways Adidas is Using AI - A Comprehensive Case Study [2025] - DigitalDefynd*. DigitalDefynd. https://digitaldefynd.com/IQ/adidas-using-ai-case-study/
Dissanayake, D. G. K., & Weerasinghe, D. (2021). Towards circular economy in fashion: Review of strategies, barriers and enablers. *Circular Economy and Sustainability*, 1–21.
Djossouvi, A. C. D., Luo, B., Sayibu, M., Debongo, D. Y. S., & Rauf, A. (2024). Social attitudes in sustainable fashion: The role of environmental knowledge on sustainable development goals (SDGs). *Journal of Advances in Management Research, 21*(4), 649–666.
Dragomir, V. D., & Dumitru, M. (2022). Practical solutions for circular business models in the fashion industry. *Cleaner Logistics and Supply Chain, 4*, 100040.
Duh, H. I., Yu, H., Venter de Villiers, M., Steffek, V., & Shao, D. (2024). Determinants of young adults' slow fashion attitudes and idea adoption intentions in Canada, China and South Africa. *International Journal of Emerging Markets*.

Durrani, S., Sohail, M., & Rana, M. W. (2023). The influence of shopping motivation on sustainable consumption: A study related to eco-friendly apparel. *Journal of Social Sciences Review, 3*(2), 248–268.

Dwikesumasari, P. R., Moslehpour, M., Sulistiawan, J., & Rizaldy, H. (2024). Investigating internal motivation in sustainable fashion consumption: Attitude towards recycled and upcycled products. *International Journal of Fashion Design, Technology and Education*, 1–13.

Echegaray, F., & Hansstein, F. (2020). Share a ride, rent a tool, swap used goods, change the world? Motivations to engage in collaborative consumption in Brazil. *Local Environment, 25*(11–12), 891–906.

Eckert, E., & Kovalevska, O. (2021). Sustainability in the European Union: Analyzing the discourse of the European green deal. *Journal of Risk and Financial Management, 14*(2), 80.

Eileen Fischer. (n.d.). *Brand Overview | EILEEN FISHER*. https://www.eileenfisher.com/our-brand.html

Elli, T., Spagnoli, A., & Iannilli, V. M. (2024). Mapping service-based retailing to improve sustainability practices in the fashion industry. *Sustainability, 16*(17), 7543.

Elmanadily, D., & El-Deeb, S. (2022). A study on consumers' purchase intentions and willingness to pay more towards eco-fashion clothes. *South Asian Journal of Management, 29*(3).

Ersoy, A. B., & Fu, S. (2021). The effect of sustainable marketing on millennials' attitudes and behavior in the fast fashion industry. *American International Journal of Business Management, 4*(6), 79–98.

European Parliament. (n.d.). *EU Strategy for sustainable and circular textiles | Legislative Train Schedule*. Retrieved February 5, 2025, from https://www.europarl.europa.eu/legislative-train/theme-a-european-green-deal/file-eu-textiles-strategy

EU strategy for sustainable and circular textiles. (2024, December 10). Environment. Retrieved December 21, 2024, from https://environment.ec.europa.eu/strategy/textiles-strategy_en

Fabiana, S., Valerio, M., Anna, P., & Mario, T. (2024). Fashion and sustainability: Evidence from the consumption of second-hand clothes. *Corporate Social Responsibility and Environmental Management, 32*, 942–967.

Faludi, J. (2022). Consciou eco-consumers or mianstram fashionistas? The perception of barriers to the ethical consumption of fashion consumers groups. *Todas as Artes, 5*(2), 10.

Farooqi, R., & Frooghi, R. (2024). Determinants of sustainable consumption behavior: A case green apparel Brands of Pakistan. *Global Management Journal for Academic & Corporate Studies, 14*(1), 77–107.

Farzin, M., Shababi, H., Shirchi Sasi, G., Sadeghi, M., & Makvandi, R. (2023). The determinants of eco-fashion purchase intention and willingness to pay. *Spanish Journal of Marketing-ESIC, 27*(3), 348–366.

Fernández, C. R. (2022, June 22). These Spider Silk Sneakers from Adidas are Biodegradable. *Labiotech.eu*. https://www.labiotech.eu/trends-news/adidas-amsilk-biodegradable-sneakers/

Filippa-K. (n.d.). *Who We Are | Filippa K*. https://www.filippa-k.com/tr/en/fk-heritage.html

Floriano, M. D. P., & Matos, C. A. D. (2022). Understanding Brazilians' intentions in consuming sustainable fashion. *BBR. Brazilian Business Review, 19*(5), 525–545.

Fung, Y. N., Choi, T. M., & Liu, R. (2020). Sustainable planning strategies in supply chain systems: Proposal and applications with a real case study in fashion. *Production Planning & Control, 31*(11–12), 883–902.

GANNI. (n.d.). *Tailoring, clothes alterations & repairs service | GANNI UK*. Ganni a/S. https://www.ganni.com/en-gb/responsibility/tailoring-repairs.html?srsltid=AfmBOoon6GoXJ1AAfYczTp5OiorX8sXs4jzN49SCdddM5Gxnx-yaZrKR

Gao, J., Zheng, P., Jia, Y., Chen, H., Mao, Y., Chen, S., Wang, Y., Fu, H., & Dai, J. (2020). Mental health problems and social media exposure during COVID-19 outbreak. *PLoS One, 15*(4), 1–10.

Garcia-Torres, S., Rey-Garcia, M., Sáenz, J., & Seuring, S. (2022). Traceability and transparency for sustainable fashion-apparel supply chains. *Journal of Fashion Marketing and Management: An International Journal, 26*(2), 344–364.

Gomes, G. M., Moreira, N., Bouman, T., Ometto, A. R., & Van der Werff, E. (2022). Towards circular economy for more sustainable apparel consumption: Testing the value-belief-norm theory in Brazil and in The Netherlands. *Sustainability, 14*(2), 618.

Guan, T. H., & Noor, S. M. (2021). Online sustainable fashion brands: Millennials' attitude, purchase intention and behaviour. *SEARCH Journal of Media and Communication Research (SEARCH), 141.*

Gupta, M., Ujjawal, N. U., & Sharma, S. (2023). Factor influencing purchase intension of sustainable apparels among millennials. *Sustainability, Agri, Food and Environmental Research-Discontinued, 12*(1).

Gurova, O. (2024). Practice theory approach to Gen Z's sustainable clothing consumption in Finland. *Young Consumers, 25*(3), 289–307.

Hahn, J. (2021, March 18). *Hermès creates Mycelium version of its classic leather Victoria Bag*. Dezeen. https://www.dezeen.com/2021/03/18/hermes-mycelium-leather-victoria-bag-mycoworks/

Hair, J. F., et al. (1998). *Multivariate data analysis* (p. 664). Pearson.

Han, H., Hsu, L. T. J., & Sheu, C. (2010). Application of the theory of planned behavior to green hotel choice: Testing the effect of environmental friendly activities. *Tourism Management, 31*(3), 325–334.

Han, J., Woodside, A. G., & Ko, E. (2024). Does consumer knowledge about sustainable-fashion impact intention-to-buy? *Asia Pacific Journal of Marketing and Logistics, 36*(10), 2390–2410.

Hansen, T. (2008). Consumer values, the theory of planned behaviour and online grocery shopping. *International Journal of Consumer Studies, 32*(2), 128–137.

Hasbullah, N. N., Kiflee, A. K. R., Roni, M., Idris, N. H., Malan, I. N. B., Jayeola, O., et al. (2022). Parasocial relationship influence on the upcycled fashion purchase intention among igeneration in Malaysia post-COVID-19: An empirical study. *Journal of Theoretical and Applied Information Technology, 100*(19), 5813–5829.

Henninger, C. E., Brydges, T., Iran, S., & Vladimirova, K. (2021). Collaborative fashion consumption–a synthesis and future research agenda. *Journal of Cleaner Production, 319*, 128648.

Herjanto, H., Amin, M., Purinton, E., & Lindle, E. L., II. (2023). Secondhand clothing purchase intentions: Generation Z's perspective. *Journal of Global Responsibility, 15*(1), 53–72.

Herold, P. I., & Prokop, D. (2023). Is fast fashion finally out of season? Rental clothing schemes as a sustainable and affordable alternative to fast fashion. *Geoforum, 146*, 103873.

Ho, S. (2022, January 4). Pangaia adds 100% biobased plant and fruit fibres to its sustainable materials menu. *Green Queen*. https://www.greenqueen.com.hk/pangaia-frutfiber-plntfiber/

Ho, T. T. H., et al. (2020). Determinants influencing consumers purchasing intention for sustainable fashion: Evidence from Ho Chi Minh City. *The Journal of Asian Finance, Economics and Business, 7*(11), 977–986.

Hoang, D. P., Nguyen, V. D. H., Chu, Q. T., & Hoang, L. B. (2022). Factors affecting behavioral and psychological perspective of Young Vietnamese customers in buying second-hand clothes. *Journal of Economics, Finance and Management Studies*. https://doi.org/10.47191/jefms/v5-i5-13

Hofstede, G. (2001). *Culture's consequences: Comparing values, behaviors, institutions and organizations across nations*. Sage.

Hohn, M. M., & Durach, C. F. (2021). Additive manufacturing in the apparel supply chain—Impact on supply chain governance and social sustainability. *International Journal of Operations & Production Management, 41*(7), 1035-1059.

Hong, Y., Al Mamun, A., Yang, Q., & Masukujjaman, M. (2024). Predicting sustainable fashion consumption intentions and practices. *Scientific Reports, 14*(1), 1706.

House of Sunny. (n.d.). *Responsibility | House of Sunny*. https://houseofsunny.com/pages/responsibility

Hu, L. T., & Bentler, P. M. (1999). Cutoff criteria for fit indexes in covariance structure analysis: Conventional criteria versus new alternatives. *Structural equation modeling: a multidisciplinary journal, 6*(1), 1–55.

Hur, E., & Faragher-Siddall, E. (2022). Young consumer perspectives on government policy interventions for sustainable fashion consumption in the UK. *Fashion Practice, 14*(3), 405–427.

Hussain, T. (2023, October 7). *10 sustainable product ideas from banana fibers*. The Textile Think Tank. https://thetextilethinktank.org/10-sustainable-product-ideas-from-banana-fibers/

Inditex. (2022, December 12). *Inditex Infinna™*. INDITEX. https://www.inditex.com/itxcomweb/tr/en/press/news-detail/d2b47ac6-a309-49a2-a79e-a380478be760/inditex-signs-a-three-year-commitment-to-buy-infinna-upcycled-textile-fiber-for-over-100-million

Islam, I. B., & Sultana, T. (2020). Factors influencing Young consumers purchase intention towards Biodegrable clothing in Bangladesh. *International Journal of Science and Research, 9*(10).

Islam, M. M., Perry, P., & Gill, S. (2021). Mapping environmentally sustainable practices in textiles, apparel and fashion industries: A systematic literature review. *Journal of Fashion Marketing and Management: An International Journal, 25*(2), 331–353.

Jian, Y., Yu, I. Y., Yang, M. X., & Zeng, K. J. (2020). The impacts of fear and uncertainty of Covid-19 on environmental concern, brand trust, and behavioral intentions toward green hotels. *Sustainability, 12*(20), 8688.

Jin, X., & Omar, A. (2023). Extending the theory of reasoned action to understand the factors influence purchase sustainable apparel intention among Chinese consumer. *Advances in Natural and Applied Sciences, 15*(3), 14–25.

Joo, K., Lee, J., & Hwang, J. (2022). NAM and TPB approach to consumers' decision-making framework in the context of indoor smart farm restaurants. *International Journal of Environmental Research and Public Health, 19*(21), 14604.

Jung, H. J., Choi, Y. J., & Oh, K. W. (2020). Influencing factors of Chinese consumers' purchase intention to sustainable apparel products: Exploring consumer "attitude–behavioral intention" gap. *Sustainability, 12*(5), 1770.

Jung, H. J., Oh, K. W., & Kim, H. M. (2021). Country differences in determinants of behavioral intention towards sustainable apparel products. *Sustainability, 13*(2), 558.

Kala, D., & Chaubey, D. S. (2024). Exploring the determinants of fashion clothing rental consumption among young Indians using the extended theory of reasoned action. *Global Knowledge, Memory and Communication.*

Kalia, N., & Kaur, R. (2021). A study on gradual increase in youth awareness towards sustainable fashion and ecological consciousness of purchase behaviour. *International Journal of Multidisciplinary Innovative Research, 3*, 1–8.

Karaosman, H., & Marshall, D. (2023). Impact pathways: Just transition in fashion operations and supply chain management. *International Journal of Operations & Production Management, 43*(13), 226–237.

Kautish, P., & Khare, A. (2022). Antecedents of sustainable fashion apparel purchase behavior. *Journal of Consumer Marketing, 39*(5), 475–487.

Khare, A. (2023). Green apparel buying: Role of past behavior, knowledge and peer influence in the assessment of green apparel perceived benefits. *Journal of International Consumer Marketing, 35*(1), 109–125.

Khare, A., & Kautish, P. (2022). Antecedents to green apparel purchase behavior of Indian consumers. *Journal of Global Scholars of Marketing Science, 32*(2), 222–251.

Khare, A., Sadachar, A., & Chakraborty, S. (2022). Influence of celebrities and online communities on Indian consumers' green clothing involvement and purchase behavior. *Journal of Fashion Marketing and Management: An International Journal, 26*(4), 676–699.

Khare, A., Sadachar, A., & Manchiraju, S. (2020). Investigating the role of knowledge, materialism, product availability, and involvement in predicting the organic clothing purchase behavior of consumers in the Indian market. *Journal of International Consumer Marketing, 32*(3), 228–242.

Kim, N. L., Woo, H., & Ramkumar, B. (2021). The role of product history in consumer response to online second-hand clothing retail service based on circular fashion. *Journal of Retailing and Consumer Services, 60*, 102457.

Kim, Y. (2023). A study of the integrated model with norm activation model and theory of planned behavior: Applying the green hotel's corporate social responsibilities. *Sustainability, 15*(5), 4680.

Kleinman, S. (2022, June 16). *How The North Face is Using a Strong Data-Based Foundation to Create the Best Customer Experience, with Sarah Kleinman, Vice President of Digital Experience - Mission.* Mission. https://mission.org/up-next-in-commerce/how-the-north-face-is-using-a-strong-data-based-foundation-to-create-the-best-customer-experience-with-sarah-kleinmanvice-president-of-digital-experience/

Kline, R. B. (2005). Principles and practice of structural equation modeling. NewYork: Guilford publications.

Koay, K. Y., Cheah, C. W., & Lom, H. S. (2022). An integrated model of consumers' intention to buy second-hand clothing. *International Journal of Retail & Distribution Management, 50*(11), 1358–1377.

Koch, J., Frommeyer, B., & Schewe, G. (2020). Online shopping motives during the COVID-19 pandemic—Lessons from the crisis. *Sustainability, 12*(24), 10247.

Kovacs, I. (2021). Perceptions and attitudes of generation z consumers towards sustainable clothing: Managerial implications based on a summative content analysis. *Polish Journal of Management Studies, 23*, 257.

Kumar, A., Prakash, G., & Kumar, G. (2021). Does environmentally responsible purchase intention matter for consumers? A predictive sustainable model developed through an empirical study. *Journal of Retailing and Consumer Services, 58*, 102270.

Kumar, N., & Mohan, D. (2021). Sustainable apparel purchase intention: Collectivist cultural orientation and price sensitivity in extended TPB model. *Journal of Revenue and Pricing Management, 20*, 149–161.

Lang, C., & Armstrong, C. M. J. (2018). Fashion leadership and intention toward clothing product-service retail models. *Journal of Fashion Marketing and Management: An International Journal, 22*(4), 571–587.

Lang, C., & Zhang, R. (2024). Motivators of circular fashion: The antecedents of Chinese consumers' fashion renting intentions. *Sustainability, 16*(5), 2184.

Leclercq-Machado, L., Alvarez-Risco, A., Gómez-Prado, R., Cuya-Velásquez, B. B., Esquerre-Botton, S., Morales-Ríos, F., et al. (2022a). Sustainable fashion and consumption patterns in peru: An environmental-attitude-intention-behavior analysis. *Sustainability, 14*(16), 9965.

Leclercq-Machado, L., Alvarez-Risco, A., García-Ibarra, V., Esquerre-Botton, S., Morales-Ríos, F., Anderson-Seminario, M. D. L. M., et al. (2022b). Consumer patterns of sustainable clothing based on theory of reasoned action: Evidence from Ecuador. *Sustainability, 14*(22), 14737.

Lee, S. E., & Lee, K. H. (2024). Environmentally sustainable fashion and conspicuous behavior. *Humanities and Social Sciences Communications, 11*(1), 1–10.

Levi's. (n.d.). *WellThread LP modelleri.* Levi's TR. https://www.levis.com.tr/wellthread?srsltid=AfmBOoqLFkaKsswo8qjclIEPJmrdst4cAgeXrlyRr1jIYFVBzo_HWH56

Liang, J., Li, J., & Lei, Q. (2022). Exploring the influence of environmental values on green consumption behavior of apparel: A chain multiple mediation model among Chinese generation Z. *Sustainability, 14*(19), 12850.

Li, L., Ming, H., Yang, R., & Luo, X. (2020). The impact of policy factors and users' awareness on electricity-saving behaviors: From the perspective of habits and investment. *Sustainability, 12*(12), 4815.

Li-Hsun, P., & Jerusalem, M. A. (2021). Study of young Indonesian's psychological attitude and purchase intention toward slow fashion. *Revista Argentina de Clínica Psicológica, 30*(1), 553.

Lim, I., & Lady, L. (2023). Factors that influence green purchase behavior by green purchase intention on green apparel. *International Journal of Indonesian Business Review, 2*(1), 1–17.

Lin, C. A., Wang, X., & Yang, Y. (2023). Sustainable apparel consumption: Personal norms, CSR expectations, and hedonic vs. utilitarian shopping value. *Sustainability, 15*(11), 9116.

Lin, P. H., & Chen, W. H. (2022). Factors that influence consumers' sustainable apparel purchase intention: The moderating effect of generational cohorts. *Sustainability, 14*(14), 8950.

Lira, J. S. D., & Costa, M. F. D. (2022). Theory of planned behavior, ethics and intention of conscious consumption in Slow Fashion Consumption. *Journal of Fashion Marketing and Management: An International Journal, 26*(5), 905–925.

Liu, L., Zhang, W., Li, H., & Zheng, Z. (2024). The impact of environmental accidents on the green apparel purchase behavior of Generation Z. *Frontiers in Psychology, 15*, 1338702.

Liu, Y., Liu, M. T., Perez, A., Chan, W., Collado, J., & Mo, Z. (2021). The importance of knowledge and trust for ethical fashion consumption. *Asia Pacific Journal of Marketing and Logistics, 33*(5), 1175–1194.

Ma, R., Yang, W., & Ketkaew, C. (2024). Young women's sustainable fashion consumption in China: A qualitative study using the theory of planned behavior. *Journal of Infrastructure, Policy and Development, 8*(9), 8135.

Maichum, K., Parichatnon, S., & Peng, K. C. (2016). Application of the extended theory of planned behavior model to investigate purchase intention of green products among Thai consumers. *Sustainability, 8*(10), 1077.

Mathur, S. (2024). Green threads: Analysing generation z's preferences for sustainable apparel. *Library Progress International, 44*(3), 7598–7612.

Mayfield, S., Gunawan, N., Burkart, M., Skip Pomeroy, R., Schreiman, A., Simkovsky, R., Samoylov, A., Bemis, T., Tessman, M., Neelakantan, N., & Mayfield, C. (2020). All-new SOLEICTM biodegradable polyurethane foams pass a critical test. In *Peer-reviewed science paper* [Press release]. Algenesis. https://cdn.prod.website-files.com/66c3ce2553273ac3f741b4a1/67d9d70249852829a562de25_5f3044e4eb0dcb547f2c98ac_Algenesis_PR%20Release_8.10.20.pdf

McCoy, L., Wang, Y. T., & Chi, T. (2021). Why is collaborative apparel consumption gaining popularity? An empirical study of US gen Z consumers. *Sustainability, 13*(15), 8360.

McMaster, M., Nettleton, C., Tom, C., Xu, B., Cao, C., & Qiao, P. (2020). Risk management: Rethinking fashion supply chain management for multinational corporations in light of the COVID-19 outbreak. *Journal of Risk and Financial Management, 13*(8), 173.

Moodly, C., Christie, L., & Strydom, M. (2023). Slow fashion consumption in Johannesburg, South Africa: Perspectives from millennial consumers. *Journal of Consumer Sciences, 51*.

Mukendi, A., Davies, I., Glozer, S., & McDonagh, P. (2020). Sustainable fashion: Current and future research directions. *European Journal of Marketing, 54*(11), 2873–2909.

Munir, S. (2020). Eco-fashion adoption in the UAE: Understanding consumer barriers and motivational factors. *Fashion Practice, 12*(3), 371–393.

Muraguri, E. K., Jin, S., & Samake, A. (2020). Modeling the role of perceived green value and consumer innovativeness in green productsâ€™ consumption intention within the theory of planned behavior. *Sustinere: Journal of Environment and Sustainability, 4*(2), 94–116.

Musova, Z., Musa, H., Drugdova, J., Lazaroiu, G., & Alayasa, J. (2021). Consumer attitudes towards new circular models in the fashion industry. *Journal of Competitiveness, 13*(3), 111.

Network. (n.d.). *N-Green - Network.* https://www.network.com.tr/content/n-green#:~:text=N%2DGreen%2C%20NetWork'%C3%BCn,%C3%A7evrenin%20korunmas%C4%B1na%20katk%C4%B1da%20bulunmay%C4%B1%20hedefliyor

Nadyasworo, A. B., Manalu, S. R., & Yusriana, A. (2023). The correlation between awareness of sustainable fashion and willingness to pay with intention to purchase sustainable fashion products. *Interaksi Online, 11*(4), 277–284.

Natural Dyes. (n.d.). *Sustainability*. https://www.naturaldyes.com.tr/sustainability/

Neupane, S., Chimhundu, R., & Kong, E. (2021). Strategic profile for positioning eco-apparel among mainstream apparel consumers. *Journal of Global Fashion Marketing, 12*(3), 229–244.

Ng, Q. X., De Deyn, M., Loke, W., & Chan, H. W. (2020). A framework to deal with uncertainty in the age of Covid-19. *Asian Journal of Psychiatry, 54*, 102263.

Nicolau, L. G. F., Christino, J. M. M., Cardozo, É. A. A., & Ferreira, F. L. (2024). The role of eco-shame in shaping behavioural purchase intentions toward sustainable fashion consumption. *Journal of Fashion Marketing and Management: An International Journal.*

Niinimäki, K., Peters, G., Dahlbo, H., Perry, P., Rissanen, T., & Gwilt, A. (2020). The environmental price of fast fashion. *Nature Reviews Earth & Environment, 1*(4), 189–200.

Nike. (n.d.-a). *Nike forward: A little care goes a long way.* Nike.com. https://www.nike.com/a/nike-forward-product-care

Nike. (n.d.-b). *Nike move to zero climate change initiative.* NIKE, Inc. https://about.nike.com/en/newsroom/releases/nike-move-to-zero-climate-change-initiative

Nudie Jeans Co. (2020, February 1). *How to wash a pair of dry jeans – Nudie Jeans | Create tomorrow's vintage | Organic cotton.* https://www.nudiejeans.com/blog/how-to-wash-dry-jeans?srsltid=AfmBOoqjh7CGaeh6UbS4qmstph30hCW0_wS7vPC0pP0JXoluY8UnjP3L

Nudie Jeans Co. (n.d.). *The Nudie Jeans Philosophy – Nudie Jeans | Create tomorrow's vintage | Organic cotton.* https://www.nudiejeans.com/info/about

Oguntegbe, K. F., Di Paola, N., & Vona, R. (2023). Communicating responsible management and the role of blockchain technology: Social media analytics for the luxury fashion supply chain. *The TQM Journal, 35*(2), 446–469.

Okur, N., Saricam, C., Iri, A. R., & Sari, I. (2023). Analyzing the impact of Covid-19 on sustainable fashion consumption with a model based on consumer value perceptions. *Journal of Fashion Marketing and Management: An International Journal, 27*, 1–25.

Ong, A. K. S., Cleofas, M. A., Prasetyo, Y. T., Chuenyindee, T., Young, M. N., Diaz, J. F. T., et al. (2021). Consumer behavior in clothing industry and its relationship with open innovation dynamics during the COVID-19 pandemic. *Journal of Open Innovation: Technology, Market, and Complexity, 7*(4), 211.

Ou, W., Tajuddin, R. M., & Shariff, S. M. (2023). *Exploring the relationship between environmental knowledge and behavioural intentions for sustainable fashion practices: Fashion design student perspectives.*

Outerknown. (n.d.). *Sustainability.* https://www.outerknown.com/pages/sustainability-page

Palomo-Domínguez, I., Elías-Zambrano, R., & Álvarez-Rodríguez, V. (2023). Gen Z's motivations towards sustainable fashion and eco-friendly brand attributes: The case of Vinted. *Sustainability, 15*(11), 8753.

Pandey, M., & Yadav, P. S. (2023). Understanding the role of individual concerns, attitude, and perceived value in green apparel purchase intention; the mediating effect of consumer involvement and moderating role of generation Z&Y. *Cleaner and Responsible Consumption, 9*, 100120.

Pangaia. (n.d.). *Planet.* https://pangaia.com/pages/impact-planet?srsltid=AfmBOopr5aI3fDU-ZUZJywOR05nPAcw5NyEPo7Q6K7_qetpNFQIiyL6K

Paras, M. K., & Pal, R. (2024). Sustainable supply chain and logistics of fashion business. In *Sustainable innovations in the textile industry* (pp. 457–474). Woodhead Publishing.

Patov, A. (2024, September 6). How Patagonia Enhances Customer Experience (CX) with a Commitment to Sustainability. *RENASCENCE.* https://www.renascence.io/journal/how-patagonia-enhances-customer-experience-cx-with-a-commitment-to-sustainability

Penz, E., & Drewes, K. L. (2022). What shapes pro-environmental attitudes and intention for sustainable fashion consumption during a stressful time event? *Sustainability, 14*(22), 15331.

QMILK. (2018, May 16). *The material of the future.* www.qmilkfiber.eu. https://www.qmilkfiber.eu/?lang=en

Rahman, O., & Koszewska, M. (2020). A study of consumer choice between sustainable and non-sustainable apparel cues in Poland. *Journal of Fashion Marketing and Management: An International Journal, 24*(2), 213–234.

Ralph Lauren. (n.d.). *Timeless by design.* RalphLauren. https://www.ralphlauren.co.uk/en/sustainability-citizenship-hub/70702272?srsltid=AfmBOorMqWpdPLN2cbRbF7N5VBxxSv1T6FhsoqsvxqaRczr8LGiN5-93

Rasmussen, N. (2020). 'Clothes' the loop: Raising awareness of sustainable fashion among millennial consumers through digital platforms. *The Boller Review, 5.*

Rastegari, H., Petrescu, D. C., & Petrescu-Mag, R. M. (2023). Factors affecting retailers' fruit waste management: Behavior analysis using the theory of planned behavior and norm activation model. *Environmental Development, 47*, 100913.

Rausch, T. M., & Kopplin, C. S. (2021). Bridge the gap: Consumers' purchase intention and behavior regarding sustainable clothing. *Journal of Cleaner Production, 278*, 123882.

RE/DONE. (n.d.). *Vintage Levi's collection*. https://shopredone.com/collections/re-done-levis

Reformation. (2023). *Ref's Circular Design Guide*.

Rinaldi, B., Ciptaningsih, E. M. S. S., & Andersean, D. E. (2024). The influence of altruism, environmental concern, and social influence on purchase intention for eco-friendly clothing products: Study of generation Z in greater Jakarta. In *IOP conference series: Earth and environmental science* (Vol. 1324, p. 012075). IOP Publishing.

Rinaldi, M., & Bottani, E. (2023). How did COVID-19 affect logistics and supply chain processes? Immediate, short and medium-term evidence from some industrial fields of Italy. *International Journal of Production Economics, 262*, 108915.

Roberts, J. (2022, August 19). *Patagonia Goes Round In Circles With INFINNA Fiber T-shirt*. FUTUREVVORLD. https://futurevvorld.com/fashion/patagonia-tee-cycle-t-shirt-infinna-fiber-infinited-circular-design-recycled-take-back-program/

Rūtelionė, A., & Bhutto, M. Y. (2024). Exploring the psychological benefits of green apparel and its influence on attitude, intention and behavior among generation Z: A serial multiple mediation study applying the stimulus–organism–response model. *Journal of Fashion Marketing and Management: An International Journal*.

Sadiq, M., Bharti, K., Adil, M., & Singh, R. (2021). Why do consumers buy green apparel? The role of dispositional traits, environmental orientation, environmental knowledge, and monetary incentive. *Journal of Retailing and Consumer Services, 62*, 102643.

Saeed, A., Sadaf, S., & Hassan, K. (2023). Creating awareness among consumers on purchase of sustainable textile products in Pakistan. *Human Nature Journal of Social Sciences, 4*(1), 360–372.

Sajjad, Z., Khalid, F., Munir, H., Arshad, S. R., & Waseem, F. (2021). Analysis of customers' perception towards second-hand garments in Pakistan's apparel industry: A qualitative study. *Journal of Innovative Research in Management Sciences*, 37–49.

Sarasi, V., Primiana, I., Harsanto, B., & Satyakti, Y. (2023). Sustainable supply chain of Indonesia's textile & apparel industry: Opportunities and challenges. *Research Journal of Textile and Apparel*. (ahead-of-print).

Selloni, D., & Selloni, D. (2017). New forms of economies: Sharing economy, collaborative consumption, peer-to-peer economy. *Codesign for Public-Interest Services*, 15–26.

Seock, Y. K., Shin, J., & Yoon, Y. (2024). Embracing environmental sustainability consciousness as a catalyst for slow fashion adoption. *Sustainable Development, 32*, 4071.

Severo, E. A., De Guimarães, J. C. F., & Dellarmelin, M. L. (2021). Impact of the COVID-19 pandemic on environmental awareness, sustainable consumption and social responsibility: Evidence from generations in Brazil and Portugal. *Journal of Cleaner Production, 286*, 124947.

Sharma, D., & Pal, H. (2020). Sustainable apparels: Examining the influence of environmental concern and environmental textile knowledge on attitude. *Journal of Textile & Apparel Technology & Management (JTATM), 11*(4).

Sharma, R., Sharma, S., & Singh, G. (2024). Exploring customer engagement on Facebook marketplace for second-hand clothing. *Social Responsibility Journal, 20*(10), 2149–2173.

Shehneh, M., Kachkar, O., & Salman, G. G. (2024). *Sustainable fashion consumption: The narrative documented from fast fashion brands and factors influencing consumers' purchase intentions in Türkiye*.

Shen, D. (2024). An empirical study on consumer awareness of sustainable fashion: A longitudinal approach. *The International Journal of Sustainability in Economic, Social and Cultural Context, 20*(1), 19.

Shin, Y. H., Im, J., Jung, S. E., & Severt, K. (2018). The theory of planned behavior and the norm activation model approach to consumer behavior regarding organic menus. *International Journal of Hospitality Management, 69*, 21–29.

Shin, S., & Lim, Y. (2021). A study on consumer confusion, value, and price sensitivity of eco-friendly fashion product. *The Research Journal of the Costume Culture, 29*(1), 48–64.

Siaw, S. D., Opoku, M., & Baiden, S. (2024). The impact of slow and circular fashion on consumers' choice of clothing apparel. *SIBATIK Journal: Jurnal Ilmiah Bidang Sosial, Ekonomi, Budaya, Teknologi, Dan Pendidikan, 3*(4), 413–436.

Silva, S. C., Santos, A., Duarte, P., & Vlačić, B. (2021). The role of social embarrassment, sustainability, familiarity and perception of hygiene in second-hand clothing purchase experience. *International Journal of Retail & Distribution Management, 49*(6), 717–734.

Sinha, P., Sharma, M., & Agrawal, R. (2022). A systematic review and future research agenda for sustainable fashion in the apparel industry. *Benchmarking: An International Journal.* (ahead-of-print).

Šimek, P., & Sadílek, T. (2024). Generation Z fashion shopping behaviour during and after the COVID-19 pandemic: The effect of sustainable consumption. *Economic Research-Ekonomska Istraživanja, 37*(1), 2341222.

Sobuj, M., Khan, A. M., Habib, M. A., & Islam, M. M. (2021). Factors influencing eco-friendly apparel purchase behavior of Bangladeshi young consumers: Case study. *Research Journal of Textile and Apparel, 25*(2), 139–157.

Sonnenberg, N. C., Stols, M. J., Taljaard-Swart, H., & Marx-Pienaar, N. J. M. M. (2022). Apparel disposal in the south African emerging market context: Exploring female consumers' motivation and intent to donate post-consumer textile waste. *Resources, Conservation and Recycling, 182*, 106311.

Spiber.inc. (n.d.). https://spiber.inc/en/protein

Szpilko, D., & Ejdys, J. (2022). European Green Deal–research directions. A systematic literature review. *Ekonomia i Środowisko, 2*, 8–38.

Stancu, C. M., Grønhøj, A., & Lähteenmäki, L. (2020). Meanings and motives for consumers' sustainable actions in the food and clothing domains. *Sustainability, 12*(24), 10400.

Stella McCartney. (n.d.-a). *BioCir® Flex: Compostable, recyclable, biobased plastic alternative | Stella McCartney UK.* https://www.stellamccartney.com/gb/en/sustainability/BioCir-Flex-Compostable-recyclable-biobased-plastic-alternative.html?srsltid=AfmBOopr-fuY4refCnXHvEcBLyfIjsnuax84Y31T3qDi0BegL6No390N

Stella McCartney. (n.d.-b). *Traceability and Blockchain technology | Stella McCartney UK.* https://www.stellamccartney.com/gb/en/sustainability/traceability-and-block-chain-technology.html?srsltid=AfmBOorWpsIL44om_l0PsUPO8B3yz8emswHR5i1sxtGnk3n34FbUFU4L

Stephan, S., & Chugh, T. (2021). Consumer sentiments on sustainability of the fashion industry: A survey. *International Journal Peer Reviewed Journal Refereed Journal Indexed Journal Impact Factor, 7*(9), 31–37.

Su, J., Iqbal, M. A., Haque, F., & Akter, M. M. K. (2023). Sustainable apparel: A perspective from Bangladesh's young consumers. *Social Responsibility Journal, 19*(9), 1654–1670.

Suhud, U., Allan, M., Wiratama, B., & Maulida, E. (2020). Slow fashion in Indonesia: Drivers and outcomes of slow fashion orientations. *Research in World Economy, 11*(6), 27–37.

Swicofil. (n.d.). *Soybean SPF - Introduction.* https://www.swicofil.com/commerce/products/soybean/152/introduction

Syahrivar, J., Kusuma, K., Pahlevi, R. A., Wei, Y., Chairy, C., & Genoveva, G. (2023). No longer look down: Investigating second-hand clothing purchase in Indonesia. *International Review on Public and Nonprofit Marketing, 20*(2), 319–339.

Tam, F. Y., & Lung, J. W. (2023). Impact of COVID-19 and innovative ideas for a sustainable fashion supply chain in the future. *Foresight, 25*(2), 225–248.

Tandon, A., Sithipolvanichgul, J., Asmi, F., Anwar, M. A., & Dhir, A. (2023). Drivers of green apparel consumption: Digging a little deeper into green apparel buying intentions. *Business Strategy and the Environment, 32*, 3997–4012.

Tang, X. (2022). Analysis on the impact of COVID-19 on the garment industries in developing countries. In *7th international conference on economy, management, law and education (EMLE 2021)* (pp. 119–123). Atlantis Press.

Tangri, K., & Yu, H. (2023). Why buy used? Motivators and barriers for re-commerce luxury fashion. *International Journal of Retail & Distribution Management, 51*(9/10), 1095–1114.

Taylor, M., White, K. M., Caughey, L., Nutter, A., & Primus, A. (2023). Unique and cheap or damaged and dirty? Young women's attitudes and image perceptions about purchasing secondhand clothing. *Sustainability, 15*(23), 16470.

Tewari, A., Mathur, S., Srivastava, S., & Gangwar, D. (2022). Examining the role of receptivity to green communication, altruism and openness to change on young consumers' intention to purchase green apparel: A multi-analytical approach. *Journal of Retailing and Consumer Services, 66*, 102938.

TheNorthFace. (n.d.). *Sustainable commitments*. Thenorthface-us.https://www.thenorthface.com/en-us/sustainability

Thorisdottir, T. S., & Johannsdottir, L. (2020). Corporate social responsibility influencing sustainability within the fashion industry. A systematic review. *Sustainability, 12*(21), 9167.

TômTex. (n.d.). *TômTex*. https://www.tomtex.com/

Tran, L. T. T. (2021). Managing the effectiveness of e-commerce platforms in a pandemic. *Journal of Retailing and Consumer Services, 58*, 102287.

Travasi, A., & Musso, F. (2023, August). A new business model for the fashion industry. The emergence of fashion-as-a-service. In *Excellence in services international conference* (pp. 126–138). Springer Nature Switzerland.

Tryphena, R., & Aram, I. A. (2023). Consumer perception on sustainable clothing among urban Indians. *Journal of Engineered Fibers and Fabrics, 18*, 15589250231168964.

Tymoshchuk, O., Lou, X., & Chi, T. (2024). Exploring determinants of second-hand apparel purchase intention and word of mouth: A stimulus–organism–response approach. *Sustainability, 16*(11), 4445.

Valenzuela-Fernández, L., Guerra-Velásquez, M., Escobar-Farfán, M., & García-Salirrosas, E. E. (2022). Influence of COVID-19 on environmental awareness, sustainable consumption, and social responsibility in Latin American countries. *Sustainability, 14*(19), 12754.

Venkatraman, P., Sayed, U., Parte, S., & Korgaonkar, S. (2021). Consumer perception of environmentally friendly antimicrobial textiles: A case study from India. *International Journal of Advanced Science and Engineering, 7*(3), 1782–1793.

Vicamara, U., Santoso, A., & Riawan, R. (2023). Thrift shopping intention: Understanding determinant of second-hand apparel shopping behavior. *Jurnal Ekonomi dan Bisnis, 26*(2), 393–412.

Vlastelica, T., Kostić-Stanković, M., Rajić, T., Krstić, J., & Obradović, T. (2023). Determinants of young adult consumers' environmentally and socially responsible apparel consumption. *Sustainability, 15*(2), 1057.

Wang, B., Fu, Y., & Li, Y. (2022). Young consumers' motivations and barriers to the purchase of second-hand clothes: An empirical study of China. *Waste Management, 143*, 157–167.

Wang, H., Gui, H., Ren, C., & Liu, G. (2021). Factors influencing urban residents' intention of garbage sorting in China: An extended TPB by integrating expectancy theory and norm activation model. *Sustainability, 13*(23), 12985.

Wang, B., Luo, W., Zhang, A., Tian, Z., & Li, Z. (2020). Blockchain-enabled circular supply chain management: A system architecture for fast fashion. *Computers in Industry, 123*, 103324.

Weekday. (2023, July 18). *Weekday*. https://www.ellenmacarthurfoundation.org/circular-examples/weekday

Woosnam, K. M., Sharma, S., Stylidis, D., & Singh, G. (2023). Understanding Fijian residents' opposition to tourism post-pandemic. *Tourism Management Perspectives, 48*, 101162.

Wren, B. (2022). Sustainable supply chain management in the fast fashion industry: A comparative study of current efforts and best practices to address the climate crisis. *Cleaner Logistics and Supply Chain, 4*, 100032.

Wu, K. J., Tseng, M. L., Yang, W. H., Ali, M. H., & Chen, X. (2023a). Re-shaping sustainable value chain model under post pandemic disruptions: A fast fashion supply chain analysis. *International Journal of Production Economics, 255*, 108704.

Wu, M., Al Mamun, A., Yang, Q., & Masud, M. M. (2023b). Modeling the reuse intention and practices of secondhand clothing: Evidence from a developing nation. *Humanities and Social Sciences Communications, 10*(1), 1–12.

Xu, J., Zhou, Y., Jiang, L., & Shen, L. (2022). Exploring sustainable fashion consumption behavior in the post-pandemic era: Changes in the antecedents of second-hand clothing-sharing in China. *Sustainability, 14*(15), 9566.

Yang, J., Al Mamun, A., Reza, M. N. H., Yang, M., & Abd Aziz, N. (2024). Predicting the significance of consumer environmental values, beliefs, and norms for sustainable fashion behaviors: The case of second-hand clothing. *Asia Pacific Management Review, 29*, 179.

Yeap, J. A., Ooi, S. K., Yapp, E. H., & Ramesh, N. (2024). Preloved is reloved: Investigating predispositions of second-hand clothing purchase on C2C platforms. *The Service Industries Journal, 44*(13–14), 993–1017.

Youn, S. Y., Lee, J. E., & Ha-Brookshire, J. (2021). Fashion consumers' channel switching behavior during the COVID-19: Protection motivation theory in the extended planned behavior framework. *Clothing and Textiles Research Journal, 39*(2), 139–156.

Zahroh, F. (2023). The determinants of sustainable apparel products in Indonesia exploring consumer "attitude-purchase intention" gap. *Return: Study of Management, Economic and Bussines, 2*(5), 502–515.

ZARA. (n.d.). *Zara provides repair service and allows for reselling and donation via "Join Life" program*. https://www.zara.com/tr/tr/z-join-life-mkt1399.html?v1=1755768

Zero waste water Technology H2 Zero. (n.d.). Jeanologia. https://www.jeanologia.com/h2zero/

Zhang, G., Wang, L., & Meng, H. (2024). Environmental knowledge level and consumer behavior regarding green fashion: A moderated mediation model. *Asia Pacific Journal of Marketing and Logistics, 36*(4), 977–992.

Zhang, Y., Liu, C., & Lyu, Y. (2023). Profiling consumers: Examination of Chinese Gen Z consumers' sustainable fashion consumption. *Sustainability, 15*(11), 8447.

Zhu, B. (2021). Millennial consumers' purchase intention for eco-fashion apparel: A study from southern China. *ABAC Journal, 41*(1), 1–25.

Digital Markets and E-Commerce Applications in Fashion Retailing

Keywords Mobile shopping applications · E-commerce · M-commerce · Consumer behavior · Technology Acceptance Model (TAM) · Theory of Planned Behavior (TPB) · Trust · Fashion retail · COVID-19

1 Introduction

As a result of technological developments and the expanding use of the internet, e-commerce has become an important sales channel in the retail industry. Online apparel sales first started in the late 1990s and early 2000s with Boo.com and Net-a-porter.com. Later, fashion products started to be sold on marketplaces such as Amazon and eBay, which are the pioneers of e-commerce and operate with both B2C and C2C business models, and then fashion retailers struggled to keep up with consumers' expanding preference for online channels throughout the early 2000s and 2010s. This type of retail has become a rapidly growing market as it allows consumers to shop regardless of time and place. Offering consumers shopping freedom, e-commerce also brings advantages such as a wide range of products, comparison opportunities, and price promotions.

For fashion brands, e-commerce allows them to appeal to wider audiences and eliminate physical limitations. The COVID-19 pandemic has accelerated the digital transformation and certainly had a major impact on the growth of fashion retail through digital platforms and applications. Due to the closure of physical stores, investments in online shopping platforms have increased. The Census Bureau's Annual Retail Trade Survey indicated that in 2020, e-commerce sales grew by 43%. When stores reopened, consumers went back to shopping in physical stores, but the share of revenue from online sales was still larger than it was before the pandemic. More than 40% of the US apparel market's income came from fashion e-commerce transactions in 2023, and the industry was estimated to be worth US $781.5 billion globally (Neilon, 2024). Many fashion brands have begun to integrate their physical

C. Saricam, N. Okur, *Changing Textile and Apparel Consumption in Transformative Era of Sustainability and Digitalization*, Sustainable Textiles: Production, Processing, Manufacturing & Chemistry,
https://doi.org/10.1007/978-3-031-97419-9_3

stores with digital platforms, combining both online and on-site shopping experiences.

The easy access of the consumer to a wide variety of products and brands through online sales has significantly changed the retail environment from business-centered to consumer-centered. A large portion of consumers actively use the internet and want to have the freedom to shop whenever they want from various electronic devices. Thus, with developing technology, companies have also started to use mobile technology applications to keep up with this change and to provide more options to the consumer.

At the point where e-commerce has reached the retail sector, with the development of various channels for shopping, other needs have emerged that go beyond offering the product that the consumer is looking for and wants from both websites and mobile shopping applications. Consumers tend to choose the products that will best meet their demands and needs. However, meeting demands and needs are not limited to product features; ensuring that the consumer has a positive shopping experience and adding value to the consumer, satisfying them, and keeping them more loyal have become priorities in e-commerce. This situation forces marketers to implement the latest trends in digital marketing methods and technologies. Understanding consumers' expectations correctly guarantees the success of these strategic efforts.

This chapter provides a broad overview of the digital markets and e-commerce applications in the fashion industry, the features of e-commerce applications including retailer websites, third-party marketplaces operating in B2C models, and mobile applications in fashion retail, and consumer tendencies specifically for mobile shopping applications, the use of which is rapidly increasing. In this context, an overview of the digital markets at global and regional levels, including market size, applications, dynamics, and consumer trends with current statistical indicators are presented. It then covers a comprehensive portrait of e-commerce applications in fashion retail, encompassing website shopping and mobile shopping, and discusses consumer characteristics, motivation, and adoption of mobile shopping applications, which have become more widespread, especially as technology becomes more prevalent in people's lives. Understanding the priorities and the expectations of the consumers is beneficial particularly for the marketers to leverage their sales and gain competitive advantage in online retail of fashion products.

The empirical study at the end of the chapter investigates the adoption of mobile shopping applications by an integrated model based on TAM (Technology Acceptance Model) and TPB (Theory of Planned Behavior) within the framework of the S-O-R (Stimuli-Organism-Response) algorithm extended with a particular factor, trust, to analyze consumer behavior regarding the adoption and future use of mobile shopping applications for purchasing fashion products. To understand the extent to which the pandemic, which has significantly affected the transformation process towards digitalization in retail, has affected consumers, the results are validated by applying a multigroup analysis among two groups of participants affected at different levels by the pandemic. Timely research is needed since consumers are evolving along with the advancement of mobile technology and the proliferation of digital shopping channels (Chi, 2018). The tendency to spend more time using mobile devices increased during the COVID-19 pandemic due to the enforcement of social distancing, lockdowns, and other measures. It is not yet known whether the effects of the COVID-19 outbreak are temporary. Thus, an empirical study is needed to understand if the preference to use mobile shopping applications is temporary, because of the fear caused by the COVID-19 pandemic. Consequently, the model and the analysis of the empirical study are expected to contribute to the academics studying in that field, and the findings may serve marketers and retailers in realizing the points to be focused on in terms of innovative content that will reach a wider range of consumers.

2 Fashion Retailing on Digital Markets

2.1 E-Commerce, M-Commerce, and Social Commerce

Fashion retailing on digital markets is the marketing of fashion products to consumers through the internet via different channels like company websites, third-party marketplaces, mobile applications, and social media. Selling products to consumers online through different means other than stores falls under the definition of e-commerce. As a subset of e-commerce, mobile commerce (m-commerce) is the buying and selling of goods and services through mobile applications on mobile devices such as smartphones and tablets. Social commerce (s-commerce) can be defined as an application of e-commerce operations and transactions in the environment of social media and social networks, and it is a combination of social and business activities (Esmaeili & Hashemi, 2019). Social media is a useful marketing device since it optimizes convenience and consumer interaction.

The diagram (Fig. 1) illustrates the relationship between online/website shopping, m-commerce, and social commerce with overlaps and differences.

Mobile applications need to be downloaded and have to be made compatible with each operating system, whereas mobile sites can just be browsed through using a web browser on any phone. Mobile sites are made to operate on smaller devices, like tablets and smartphones. The design, content, and functionalities are made compatible with mobiles so that they do not only show the shrunk version of the desktop site. For example, the mobile site of ASOS is designed to provide smooth shopping on mobile phones and tablets. It features easy navigation, mobile-optimized images, fast product pages with quick load times, and a responsive design that accommodates different screen sizes. Customers can view collections, browse product details, and make purchases through the mobile web browser itself without installing a particular application (Schroeders & Visser, 2019).

While social media accounts were initially used to promote products and lead customers to websites or mobile applications to buy, they now also allow one to

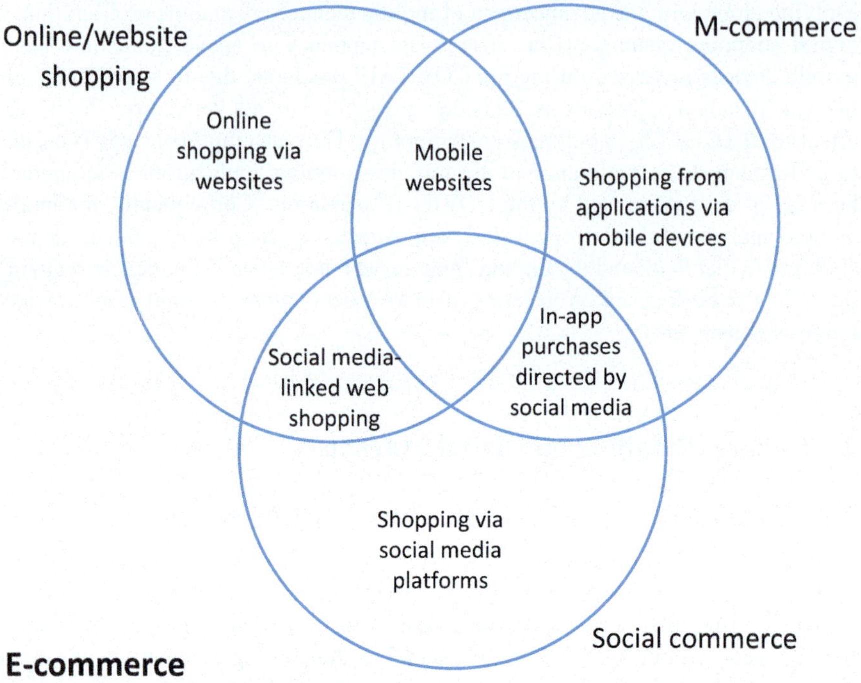

Fig. 1 Relationship between online/web shopping, m-commerce, and social commerce

shop on the social media platform itself without leaving the platform. H&M has a fully integrated Instagram Shop, where customers can shop collections, see prices, and purchase directly within the application. The brand also makes use of shoppable stories and posts on a consistent basis, where users can tap on tagged products and directly proceed to the checkout page (Salas, 2024). Louis Vuitton implemented Instagram shopping, and users can scroll through new collections, tap on product tags, and checkout within the application. The brand leverages TikTok Live Shopping, where brand ambassadors and influencers showcase goods in real-time with the ability to purchase (https://www.voguebusiness.com/story/consumers/bake-at-home-totes-and-meme-culture-the-luxury-brands-winning-digital). Nike uses Snapchat AR Shopping, where users can try on shoes virtually and shop from the application (https://forbusiness.snapchat.com/inspiration/nike-new-york-augmented-reality). Nike also has been included in Instagram in-app shopping, as many brands like Adidas, Balmain, Burberry, Prada, and Zara, which uses Instagram checkout feature allowing to proceed for payment without leaving Instagram (Lovejoy, 2019).

2.2 History and Evolution of E-Commerce in Fashion Retailing

One of the first fashion brands to become well-known for its online store was Boo.com, which was founded in 1998 and headquartered in London, England. The company targeted young, well-off, and fashion-involved consumers between the ages of 18 and 24 with the vision of being the world's largest online retailer of prestigious leisure and sportswear brands like Tommy Hilfiger, Lacoste, Polo Ralph Lauren, Nike, Fila, and Adidas. Boo.com, the first in electronic fashion retailing, actually realized today's digital transformation a quarter of a century ago; and created a digital salesperson named "Miss Boo" to guide users at each step of shopping and give helpful tips. The company also used 3D virtual technology that allowed customers to try on clothes on models. However, the high marketing and public relations expenses along with these technological investments, and the growth plan of operating simultaneously in several European countries, caused its activities to end in approximately 18 months. Problems with users' experience, such as the website working slowly and the page taking a long time to load, were difficulties encountered on the consumer side and limited the effectiveness of high-cost advertising campaigns. Besides, a high rate of product returns caused sales to fall short of expectations (Chaffey, 2014).

In the same year that Boo.com was defuncted, another London-based firm, Net-a-Porter, launched in the online fashion market. The founder Natalie Massenet, who set out to sell designer collections and high fashion brands online, has achieved success with her extensive knowledge and experience of luxury fashion brands. When Swiss luxury goods manufacturer Richemont saw that an increasing volume of its sales was coming through Net-A-Porter, it purchased a majority of the company's shares in 2010, giving it the capital needed to expand globally. After this sale, the company's value increased to US $533 million. The company has grown by diversifying its target audience with the launch of Outnet.com as an online designer fashion outlet in 2009, MrPorter.com in 2011, and finally Petite-A-Porter.com in 2013. In 2015, Net-a-Porter merged with an Italian company Yoox, which launched in 2000 and owns three global e-commerce sites: yoox.com, thecorner.com, and shoescribe.com, which sell designer collections. After the merge, Yoox Net-a-Porter leveraged its capabilities and reached a wider audience by combining Yoox's specialization in logistics and operations and Net-A-Porter's excel at interacting with customers (Roll, 2017). Yoox Net-a-Porter is still one of the world's largest luxury fashion destinations with 9 M+ social media followers, 1.9 M+ highly engaged email & push subscribers, 2.2 M+ app downloads across all platforms, and 7 M+ global site visits every month (https://www.net-a-porter.com/en-tr/).

By the early 2000s, US-based e-commerce giant Amazon expanded into a variety of products including apparel. Today, with a share of over 40%, apparel ranks second after electronics in Amazon's best-selling product categories (https://financesonline.com/amazon-statistics/). Today, all fashion brands generate high sales volumes from their websites, mobile applications and third-party marketplaces. E-commerce enables brands to reach a global audience without the need for physical stores. According to UNCTAD's Digital Economy Report 2021, cross-border B2C e-commerce sales reached approximately US $440 billion in 2019, marking a 9% increase from 2018. This growth underscores the expanding role of cross-border e-commerce in the global fashion industry, accounting for 20% of the industry's total sales in certain markets (UNCTAD Report, 2021). Global platforms like Farfetch and Zalando thrive by connecting international customers with brands and retailers worldwide. The global platforms have leveraged their extensive digital reach and personalized customer experiences to cater to a global audience, solidifying their positions in the competitive fashion e-commerce market. Moreover, e-commerce platforms support sustainable practices and circular fashion. Secondhand fashion platforms have grown by 50% year-over-year, as highlighted by ThredUp's 2023 Resale Report. Consumers value the accessibility of eco-friendly and pre-owned items through online platforms (ThredUp Resale Report, 2023).

Due to the rise in internet access and the use of smartphones, as well as the COVID-19 pandemic more recently, e-commerce has experienced exponential growth in the last decade. Thanks to mobile devices, people are constantly online, which makes purchasing even easier. Thus, mobile applications are becoming increasingly common in the retail industry. Customers' purchasing decisions have been greatly impacted by these developments, and it has forced retailers to analyze consumer behavior more deeply. For instance, consumers use online shopping channels not only for reasons such as discounts, price advantages, and product variety but also for features such as richness of information, high-resolution displays, fast response times and user-friendliness. Moreover, as e-commerce is developing in fashion retail rapidly, changing customer expectations due to increasing competition and trends have brought about a transformation in digital fashion retail. While fast delivery and ease of return are the most important parameters affecting consumer preferences, personalized shopping experience increases customer satisfaction and positively affects sales. With the developing technology, the ability to offer product recommendations according to customers' interests, shopping habits, and historical data with the help of artificial intelligence (AI) and data analytics makes e-commerce attractive for the consumer. While it can be challenging for companies to customize and tailor user experience, consumers today often choose to shop from retailers that offer personalization. Therefore, personalization is the focus of many global brands for enhancing consumer engagement and satisfaction. For instance, H&M has collaborated with Google Cloud to improve data analytics capabilities and secure using AI. Another Example is fashion giant Zara's partnership with Jetlore to improve operational efficiency and provide growth in sales. Through the use of AI for predictive consumer analytics, sales growth is expected with the help of deeper consumer insights (Sairam, 2024).

Nike, one of the world's leading global brands in sportswear, also stands out as a brand that keeps up with the digital transformation in retail as a brand that invests in AI-based technology. On the analysis of consumer behavior side, Nike employs AI for deep customer analytics, aided by its acquisition of Zodiac (Barseghian, 2022).

To assist some online buyers in creating location-based tailored website images, Puma implemented Google Cloud Imagen 2. For instance, shoppers in Japan can view a picture of trail running shoes at the base of Mount Fuji (https://www.retail-insight-network.com/news/puma-google-cloud-ai/).

Another sportswear brand that recognizes how AI can revolutionize personalization is Adidas. The company has moved beyond traditional email marketing tactics by incorporating cutting-edge analytics and has adopted a personalized and dynamic communications strategy that appeals to a wide range of consumer tastes. For Adidas, AI is key to gaining deeper insights into its customers. To gain a comprehensive understanding of each customer's distinct tastes and journey, the brand looks at browsing history, engagement patterns, and purchasing behavior in addition to traditional demographics (Keru Consulting, 2023).

When looking at the digital transformation applications of luxury fashion in retail, brands like Burberry, Gucci, and Louis Vuitton are seen redefining the customer experience, innovating product offerings, and increasing operational efficiency by leveraging AI. For instance, through innovative virtual try-on technologies, Burberry is embracing artificial intelligence by enabling clients to view clothing items on a 3D model of themselves prior to purchase, providing information about how the items will fit. Gucci's AI-powered mobile application enhances the in-store consumer experience by providing sales assistants with data about product details and past purchases. By providing AI-powered chatbots and mobile applications, Louis Vuitton has combined AI with customer support to improve the purchasing experience. These developments respond to inquiries, offer personalized product recommendations, and even make scheduling appointments easier (Behrooz, 2023).

Another area of digital transformation to remain competitive in fashion is social media marketing and social commerce. In order to reach and interact with more consumers, social media platforms have become an indispensable marketing tool and also shopping platform in fashion e-commerce in recent years. It is a fact that today's internet users have an account on at least one platform. Social media is widely used and serves different purposes. It is possible to say that some platforms have, in a sense, turned into e-commerce platforms. Therefore, social media can be seen as both a marketing tool and a gateway to e-commerce. From the consumer perspective, it is easy to get information and recommendation about the products, and purchasing in a single step. Besides, e-commerce platforms use customized communications to persuade customers to purchase, and social media enables marketers and sellers to create and distribute highly customized contents. Personalizing social media advertisings is a trending approach to increase online sales and companies that use social media platforms to personalize their products or services and encourage purchases have seen an increase in their revenues (Li & Shen, 2023; Rai & Yadav, 2022).

Platforms like Instagram, TikTok, and Pinterest drive traffic to e-commerce sites through targeted advertisings and influencer endorsements. In 2023, social media became a major shopping channel worldwide, with approximately 14% of shopping journeys beginning on these platforms. 88% of Generation Z and Millennial consumers used social media to discover new products or seek shopping inspiration.

Additionally, 44% utilized social media to evaluate items before making a purchase. In 2024, social commerce accounted for an estimated 19% of global e-commerce, and is expected to account for 22% of total e-commerce worldwide by 2028 (Dean, 2024). The "Shop Now" feature directly links users to e-commerce sites, boosting sales. Recently, social media platforms have introduced digital storefronts, enabling brands or stores to show their products, interact with customers, and facilitate purchases, all within the platform itself. Platforms like TikTok have integrated in-app shopping functionalities, enabling users to purchase products directly through the application. This integration has significantly boosted sales, as evidenced by brands like P Louise, a cosmetic brand, which sold over US $2 million worth of products during a 12-hour TikTok Live shopping event (May, 2024). These developments underscore the growing influence of social media platforms in shaping consumer purchasing behaviors, particularly among younger demographics.

Interaction between the user and the brand is the core of social media sales. In social media, content marketing and influencer marketing allow brands to establish a close bond with consumers. The goal of content marketing is to produce and distribute excellent, timely, and consistent information to draw in and hold on to audiences and eventually engage consumers and encourage user action that would likely enhance sales (Wei et al., 2024). Fashion consumers want to access and evaluate fashion trends and brands to make informed choices for potential purchases. Through content marketing, customers can actively interact with the brand and as a result, they develop a positive bond with the brand. This interaction and close relationship increase the likelihood that consumers will purchase from the brand. The trust that is developed between the brand and the consumer also leads to brand loyalty (Kamaldeen, 2024).

The increased use and exposure of technology facilitates the engagement of companies with the brands to engage with consumers through social media or other digital shopping platforms. This plays a significant role in conveying sustainability practices and, which is the primary goal of the entire world and especially the fashion industry, to the consumer (Parker & Henninger, 2018).

Fashion brands frequently collaborate with influencers in various ways such as product placements, sponsorships, and influencer-hosted events. Since discussing any topic related to products or sharing brand-related content with others is defined as word-of-mouth (WOM) marketing, influencer marketing is a type of WOM marketing that uses influencers to boost the brand to a wider audience. More precisely, since it is done in a virtual environment through social media platforms, it is possible to define influencer marketing as a type of electronic word-of-mouth (eWOM) marketing (You & He, 2023). Today, social media influencership essentially means the frequent dissemination of the contents via social media channels. Influencers are therefore individuals who have above-average potential to influence others due to attributes such as frequency of communication, personal persuasiveness, or size and centrality of a social network (Zietek, 2016). Influencers play an important role in purchasing intention because influencers have the power to direct trends and consumer preferences, and consumers perceive these recommendations as relatable and trustworthy. Brands may build vibrant communities around their products by collaborating with fashion influencers and establishing a strong social media presence.

3 E-Commerce Applications in Fashion Retail and Consumer

3.1 Global Online Shopping and Mobile Commerce Markets

The fashion e-commerce market has experienced significant growth in recent years and is projected to continue expanding rapidly. According to the Business Research Company, valued at US $888.56 billion in 2024, it is expected to reach US $974.87 billion in 2025, reflecting a compound annual growth rate (CAGR) of 9.7%. This upward trend is set to accelerate, with the market projected to hit US $1505.09 billion by 2029 at a CAGR of 11.5% (Fashion E-commerce Global Market Report, 2025).

The US fashion e-commerce market has experienced significant growth in recent years. In 2024, retail e-commerce revenue from the sale of fashion apparel, footwear, and accessories was projected to reach over US $145 billion (https://www.statista.com/topics/3481/fashion-e-commerce-in-the-united-states/). In 2023, the United Kingdom led the European Fashion e-commerce market with US $36 billion, while Germany followed with US $19.88 billion (https://www.statista.com/forecasts/1344694/fashion-e-commerce-revenue-europe-country). According to Europe Online Fashion Apparel Footwear Market report published by Research and Markets for 2024, the European fashion e-commerce market is projected to grow to about US $198.77 billion by 2028, representing a compound annual growth rate (CAGR) of 10% from 2024 to 2028 (Europe Online Fashion (Apparel, Footwear) Market 2024).

The Asian fashion e-commerce market experienced significant growth, with projections indicating a continuation of this trend in the coming years. The market was valued at approximately US $521 billion in 2023. It is anticipated to reach around US $985 billion by 2030, with a compound annual growth rate (CAGR) of 9.5% from 2023 to 2030 (https://www.coherentmi.com/industry-reports/asia-fashion-ecommerce-market/market-size).

As the largest fashion e-commerce market in Asia, China's online fashion revenues were approximately US $360.4 billion in 2023. The market is expected to continue its expansion, with projections estimating revenues to reach around US $450 billion by 2028, reflecting a CAGR of 4% from 2024 to 2028. Ranking second, South Korea's fashion e-commerce market generated revenues of about US $25.5 billion in 2023. The market is projected to grow to approximately US $29.5 billion by 2028, with a CAGR of 2.6% over the same period. In 2023, Japan's fashion e-commerce market was valued at around US $22.2 billion, making it the third-largest in Asia. However, the market is expected to experience minimal growth, with a projected CAGR of only 0.2% from 2024 to 2028. India and Indonesia are also notable, with India's online fashion revenues at approximately US $17.3 billion and Indonesia's at around US $16 billion in 2023. Indonesia, in particular, is expected to witness significant growth, with a projected CAGR of 21% from 2024 to 2028 (Fashion eCommerce in Asia: Top 5 Countries by Revenue).

The growth in global e-commerce markets is driven by factors such as the adoption of new payment methods, rising global awareness and accessibility of e-commerce, economic development, increased smartphone usage, evolving consumer preferences, rapid urbanization, and a growing demand for sustainable and ethical fashion. Notable emerging trends shaping the market include the integration of AI-powered virtual assistants, the use of augmented reality for enhanced shopping experiences, digital marketing and social media, accessing cross-border fashion commerce, and introduction of sustainable fashion collections in fashion e-commerce platforms supporting small eco-friendly brands (Fashion E-commerce Global Market Report, 2025).

3.2 *Characteristics of Online Shopping Websites and Mobile Applications*

Online shopping websites and mobile shopping applications are digital platforms that facilitate browsing, purchasing, and payment of goods and services. Sales in a virtual environment without the need for a physical store first started on websites. Online shopping allows customers to make purchases through a retailer's online store using a device that has an internet connection, usually a personal computer (Yang, 2012). Mobile shopping, on the other hand, takes place on mobile devices and allows consumers to browse and purchase via mobile devices (Shao Yeh & Li, 2009). Mobile shopping is considered as a distinct online sales channel that uses mobile devices to distribute content, communicate business information, and carry out transactions (Leung & Antypas, 2001; Zwass, 2003). Mobile shopping applications downloadable to a mobile device, extremely changed the way, which individuals interact with the surrounding environment (Milanova & Saprikis, 2018). The amount of time spent by consumers on mobile devices is steadily rising, which supports the expansion of mobile shopping.

Below are the characteristics of online shopping websites and mobile shopping applications. They have a significant role in shaping consumer attitudes.

- **User-friendly interface**: Online and mobile shopping applications prioritize intuitive design to ensure ease of navigation and use. Clean layouts with a clear categorization of products, and simplified checkout processes (e.g., one-click purchasing) can be considered as important factors in terms of user-friendliness and hence customer satisfaction (Dianat et al., 2019). Including similar features found on websites, such as navigation, search options, and share functions, mobile-friendly applications are designed to display content in a way that is easily readable and accessible on mobile devices, including smartphones and tablets (Liu & Lai, 2015). Guo et al. (2023) stated that shopping websites with excellent visual, navigational, and information design are seen as more user-friendly and enjoyable, leading to higher customer retention. While navigation and information design are key factors in determining usability, visual design plays a significant role in enhancing satisfaction. This term is defined as mobile-friendliness for mobile shopping applications.

- **Informative content and reliability of the information**: Informative content encompasses all the details and descriptions with easily readable text that helps consumers understand and evaluate products or services before making a purchase. Informative content is a critical component of online shopping platforms, influencing consumer decision-making, trust, and overall satisfaction (Magrath & McCormick, 2013).
- **Personalized recommendation**: With the help of AI and machine learning, tailored recommendations based on browsing history, purchase behavior, and preferences are offered, which has a significant role in improving consumer engagement and loyalty and leading to long-term success (Babatunde et al., 2024).
- **Multichannel integration:** Online and mobile shopping applications can integrate both offline and online shopping experiences. Real-time inventory checks, and BOPIS (Buy online, pick up in-store) options offer the consumer the opportunity to reach the product quickly and get rid of shipping costs. The effectiveness of omnichannel strategies can enhance customer loyalty (Verhoef, 2021).
- **Multiple and secure payment systems:** Offering multiple payment options (credit cards, online payment systems like PayPal, digital wallets, and Buy Now Pay Later/BNPL) are critical for gaining consumer trust. Secure and versatile payment systems drive higher conversion rates, build trust, and reduce cart abandonment (Agrawal, 2021; Solberg Söilen, 2024).
- **Information privacy and security:** These are critical factors influencing consumer trust, satisfaction, and adoption of online shopping platforms. They encompass the protection of personal and financial information, ensuring that users feel safe when engaging in digital transactions. Consumers are increasingly aware of how companies use their data, including for targeted advertising and personalized recommendations. Lack of transparency about data handling can deter customers from shopping online. Low privacy and security lead to dissatisfaction and lower purchase intention (Rodríguez-Torrico et al., 2019).
- **Consumer-led interactions:** It refers to the consumer's direct engagement with the design elements of a website or mobile applications and their interactions with the retailer (Yoon et al., 2008). Features like functional symbol keys and drop-down menus, for instance, enhance the user experience by allowing consumers to navigate and control their interactions with the retailer. Another dimension in this context is consumer reviews and opinions. Consumers no longer rely solely on advertisements or professional advice when making purchasing decisions. Instead, they primarily depend on online consumer reviews, as future buyers actively seek feedback from the reviews. Consumer-led interactions not only improve customer satisfaction (Roy Dholakia & Zhao, 2009) but also help

retailers build and strengthen buyer-seller relationships (Feinberg & Kadam, 2002; Magrath & McCormick, 2013).

- **Social media integration:** Product sharing, reviews, and ratings on social media keep the brand fresh in the consumer's mind. Consumer engagement increases when brands integrate social media with online shopping by enabling direct purchases through shoppable posts that link to their website or mobile applications on their own social media accounts or in influencer-driven campaigns. Social interaction influences consumer trust in online shopping (Liang et al., 2011).
- **Sales promotion:** Promoting products and offering discounts or campaigns to encourage purchase intentions is a common practice on online shopping sites and mobile applications. Web or application-specific discounts, and special discount codes through influencer collaborations are frequently used to increase sales (Magrath & McCormick, 2013).

In Table 1, differences between mobile devices and desktop and laptop computers are stated (Yang, 2020). Both desktop/laptop computers and mobile devices offer distinct advantages and drawbacks, making them suitable for different consumer types and usage scenarios. Desktop and laptop computers provide a superior experience in terms of screen size, efficiency, and productivity, making them ideal for work-related tasks that require full attention and active searching. However, they lack portability and require time to power on. In contrast, mobile devices offer convenience, accessibility, and sensory engagement, making them more suitable for entertainment, casual browsing, and impulsive purchasing. However, their smaller screens and fragmented usage patterns can hinder productivity and task completion. Ultimately, the choice between the two depends on the user's needs, whether for focused work or on-the-go convenience, or consumers' characteristics and lifestyle. For example, Ladhari et al. (2019) noted that Generation Y shoppers do not have similar motivations for online shopping and use different devices (e.g., desktops, laptops, and smartphones). The researchers divided the fashion shoppers into six categories (price shoppers, fashionista shoppers, emotional buyers, discovery

Table 1 Differences between mobile devices and desktop and laptop computers (Yang, 2020)

Desktop and laptop computers	Mobile devices
Greater screen size, enjoyable experience, and excellent effectiveness	Small screen, relatively challenging keyboard input, rare to complete a complex task
Turning on and using it at any time or place is impossible, and waiting for the system to switch on takes time	Able to use anytime and anywhere, sensible location, sensible sound, sensible movement, and sensible image
Since computers are mostly used for business purposes, users' computer usage time is full, productive, and compatible	Since mobile phones are used mostly during commuting or at home in a relaxed state, time fragmentation is pronounced, inconsistent, and easily interrupted
Clear demand and active search are the primary objectives, entertainment and leisure needs are secondary	While the demand for leisure and entertainment emerges as the primary goal, it also brings about impulsive purchasing demand, such as the immediate demand for sales

buyers, shopping fans, and strategic buyers) and indicated that price and discovery shoppers mostly use laptops, with nearly 75% of price shoppers using a laptop or desktop computer for online shopping, whereas emotional buyers are the most inclined to shop using mobile devices (smartphones or tablets), followed by fashionistas consumers and shopping fans. Among these groups, shopping fans are the most likely to make online purchases via smartphones.

3.3 Online Shopping Websites and Mobile Applications in Fashion and Apparel Consumption

The characteristics of online shopping websites and mobile shopping applications defined in the previous heading are also valid for online and mobile shopping applications in fashion and apparel consumption. However, some additional features should also be mentioned that are listed below:

- **Multimedia product viewing and visual appeals:** This refers to the consumer's capability to explore a retailer's online product offerings through multimedia features. These may include photography, graphics, multimedia videos, and interactive image functionalities like zoom, rotation, and catwalk videos displayed on individual product pages (Magrath & McCormick, 2013).
- **Augmented reality (AR) and virtual try-ons:** One of the additional features of online and mobile shopping applications in fashion is augmented reality (AR) and virtual try-ons. AR features allow users to try on clothes virtually or visualize how items look in real life. Add an interactive and playful element to shopping experience. Similar features are also used in the sale of cosmetic products. AR enhances consumer confidence, leading to fewer returns (Poushneh & Vasquez-Parraga, 2017). It is also identified as a key driver of consumer engagement in e-commerce (McLean & Wilson, 2019).
- **Information on sustainability practices:** Providing sustainability details related to products on the website or mobile applications or options such as filtering sustainable products are attractive to eco-conscious customers. Transparency in sustainability improves consumer trust and brand equity (Förnges & Sieling, 2023).
- **Loyalty programs**: Loyalty refers to customers' ongoing preference for consuming specific products. It reflects their commitment and demonstrates a consistent intention to repurchase preferred products or services. Reward points, badges, and exclusive discounts can be used in loyalty programs. Loyalty programs can foster emotional connections and a sense of commitment among consumers. A loyalty program has consequences for customers' satisfaction and trust (Sumiyana & Komariyah, 2024).
- **Entertainment and gamification:** Online and mobile shopping applications often incorporate engaging and entertaining elements like gamification, interactive features, and social sharing to make the shopping experience enjoyable.

Daily check-in bonuses or interactive games combine fun with functional shopping experiences. For instance, Shein's shopping website and mobile application gamify shopping with point systems (https://us.shein.com/bonus-point-program-a-371.html).

When all features of online shopping and mobile applications in fashion and apparel consumption are considered in their entirety, they can be categorized into three as utilitarian, hedonic and social features (Pop et al., 2023).

The utilitarian features include functional characteristics such as user-friendliness (ease of use and usefulness), personalization, privacy, and security. Perceived usefulness, in particular, has been identified as the strongest determinant of the perceived value of fashion mobile applications (Sun & Chi, 2018; Murillo-Zegarra et al., 2020). Additionally, consumer perceptions of privacy and security play a vital role in encouraging mobile application purchases. Lower concerns about privacy and security correlate with greater satisfaction and trust (Sun & Chi, 2018; Rodríguez-Torrico et al., 2019).

The hedonic aspects, such as entertainment, fun, visual appeal, enjoyment, and pleasure, have been identified as key factors influencing consumers' positive attitudes and their intentions to continue using an application (Chang et al., 2016; Lee & Kim, 2019).

The social features cover consumer reviews and opinions. With the rise of digitalization and the increasing prevalence of online shopping, online reviews and customer feedback have gained prominence in influencing consumer decision-making. These reviews have the potential to shape consumer attitudes (Soni et al., 2019; Doh & Hwang, 2009).

In addition, the elements that should be included in the applications to support and encourage the behavior for consumers who are looking for sustainable fashion products are stated as being glanceable, subtle, functional, symbolic, and enjoyable (Parker & Henninger, 2018). According to Hinman (2012), sustainability routes and indicators ought to be readily apparent at first glance rather than tucked away in submenus. Preece et al. (2015) suggested that indicators should be considerate of the user's duties and desired outcomes, which is unlikely to involve locating sustainable clothing unless the application is designed exclusively for an ethical or sustainable brand. Sustainability indicators should include real-world metaphors to symbolize sustainability practices that make the application has a personality such as fun, minimalist, serious, etc., and this should align with the intended impact on consumers (Yarmosh, 2011). Adding sustainable triggers/facilitators throughout the application should be a core feature, and enjoyable and aesthetically pleasing interfaces are perceived as having higher functionality. Thus, it is important to enhance both hedonic and utilitarian experiences for application usability (Parker & Henninger, 2018).

Currently, mobile shopping applications are used not only to sell new/unused products but also as platforms where second-hand clothing is offered for sale or rental options. Applications designed for this purpose should also have certain features. In such applications, it is important to emphasize the importance of reacquisition by matching it with consumers' preferences and ethical values. Besides, the benefits of reacquisition to the consumer may include economic benefits to the consumer budget. The mobile applications can support users' financial constraints by setting reasonable prices for second-hand products or offering discounts. This can positively affect the success of the second-hand mobile applications due to consumers who cannot bear the high price of purchasing sustainable products and who prefer second-hand clothing out of necessity. User engagement and providing an enjoyable and immersive shopping experience are also among the important features for second-hand mobile applications. Consumers may value transparency and authenticity, and second-hand shopping applications can provide insider information about these issues as well as their social impact. Consequently, applications which have social media integration, functionality, enjoyable and immersive shopping experience, and transparent and authentic aspects not only meet the consumer needs but also actively contribute to the positive evolution of second-hand fashion field (Johnson, 2023).

4 Use of Mobile Shopping Applications in Fashion and Apparel Consumption

4.1 Consumer Characteristics in Use of Mobile Shopping Applications

Consumer characteristics that influence the use of mobile shopping can broadly cover demographic characteristics. In this regard, the age, gender, education level, and income level of consumers are closely related to the use of mobile shopping and the preference of applications due to the features they offer or have, and consumers' purchasing behavior.

With regard to gender, in the early days of internet use, women were less likely to shop online and spent less than men, as men initially showed a stronger interest in computers. But this trend has since shifted, with women now using the internet more frequently than men, and female consumers tend to shop more frequently for clothing via mobile applications (Ruane & Wallace, 2013; Miladinovic & Hong, 2016). Studies indicate that society tends to prioritize women's physical appearance more than men's (Bakewell et al., 2006), and women are more inclined to use clothing and fashion as a means of expressing their "self-identity" and defining themselves. Since online shopping provides consumers with access to highly specific or niche products that are often difficult to find in physical stores, it enables women to meet their "self-expression" needs by exploring distinctive brand options (Ruane & Wallace, 2013).

There are also differences in fashion purchasing via digital markets according to age and generational cohorts. According to the RetailX Fashion Sector Report, based on a survey of 1571 online fashion consumers in the USA, UK, Germany,

Brazil, France, and Australia conducted in 2022, among online fashion consumers, Generation X and Generation Y (Millennials) are identified as key age groups for online shopping. Contrary to popular belief, the report states that consumers over the age of 55 are quite knowledgeable about mobile and have embraced e-commerce, and purchase fashion products online. Generation Z consumers aged 16–24 make up 30% of online fashion consumers, while shoppers over the age of 45 make up 40% of online consumers. In the report, this situation is explained with diverse income levels. Millennials and Generation X consumers are said to be the majority of middle-income and upper-income earners in most countries, and they account for 71% of online fashion consumers, as they have a large discretionary spend. On the other hand, young generations are highly proficient with technology and favor online shopping; however, their lower incomes and limited spending power restrict their purchasing capacity. As a result, despite their digital expertise, Generation Z consumers may be less inclined to invest in high-end fashion products (https://ecommercegermany.com/blog/fashion-online-shopper-insights). However, the definition of high-end is noticeable here, because access to fashion products is not only limited to high-end products on digital markets.

There are different views on the income level and purchasing power of Generation Z. Their income levels are said to be lower, especially for young individuals who are continuing their education or at the beginning of their careers. However, it is stated that Generation Z has an estimated spending power of 450 billion dollars worldwide (360 million dollars in the US alone). Almost half of this generation is now adults and has joined the workforce (Öztopçu, 2024). In addition, according to a report by Bank of America, the total income of Generation Z is expected to increase approximately fivefold by 2030, reaching 33 trillion dollars. This shows that the spending power of Generation Z will increase significantly (https://www.paradergi.com.tr/dunya/2021/09/09/z-kusagi-10-yil-sonra-hepimizden-zengin-olacak). Therefore, Generation Z consumers are not a homogeneous group and their current income level and purchasing power may vary depending on their age and career stage. However, it is expected that the economic impact and spending power of this generation will increase in the future. Besides, Generation Z makes up over 32% of the global population and represents the fastest-growing group of online consumers (Arumugam & Wing, 2020). In support of this, in another market survey, it was stated that 51% of mobile apparel consumers in the United States as of March 2023 are between the age of 18–24, whereas around 12% of consumers are at the age of 45 and above (https://www.statista.com/statistics/1329338/ clothing-shoppers-mobile-audience-united-states-age/). Addressing mobile apparel shopping specifically, data shows that mobile applications are preferred more by young consumers. In this case, social media and influencer marketing play a crucial role.

Since they grew up with modern technologies and are heavily dependent on the internet both in their daily lives and at work, Generation Y and Z are often described as tech-savvy, even described as digital natives, who are highly proficient with technology and devices and have a strong attachment to social media (Bolton et al., 2013; Wei et al., 2024). However, the belief that older generations are far from technology is a misconception. For example, a study conducted in Spain for investigating online consumer behavior within the Spanish fashion industry revealed that the penetration rate of internet use increased the most in the 65–74 age group (from 2.3% in 2006 to 65% in 2023), with the second highest increase in the 55–64 age group. This indicates that the digital divide is narrowing, with older generations increasingly integrating digital technologies into their daily lives. However, in the same study, it was found that more than 63% of those aged 55–64 have shopped online at least once, while only 39% of those aged 65–74 have shopped online at least once. On the other hand, this value is at 91% for the 25–34 age group (Martin, 2024).

While older generations may also be skilled in the use of technology, some barriers may limit mobile application adoption. Although Generation X is technologically adept, they often exhibit certain insecurities. In contrast, many aspects of mobile shopping that create uncertainty or concern for Baby Boomers and Generation X, such as delivery fees, consumer tracking, and entering credit card details, do not have the same impact on younger generations. For Millennials and Generation Z, mobile shopping is perceived as significantly less risky, with e-commerce, barcodes, and online payment technologies seen as standard components of the digital economy. Generation Z consumers have no issues with trusting mobile shopping applications. As a result, mobile shopping, which involves financial and privacy risks, is primarily suitable for individuals who can navigate these challenges due to their openness to new experiences and positive outlook on mobile shopping (Lissitsa & Kol, 2021; Vinerean et al., 2022). In a study of UK mobile fashion consumers conducted in 214, Generation X often preferred online shopping via websites instead of mobile applications due to the excess of options, while Millennials and Generation Z complained that product display problems on smartphones make it difficult for them to shop online and make judgments (Tupikovskaja-Omovie et al., 2014). This has actually revealed the importance of increasing product variety and providing ease of product display in mobile shopping applications in fashion.

In addition, tools that increase ease of use and usefulness can increase the mobile application usage of older generations. Indeed, embedded artificial intelligence-mobile smart speech recognition (AI-MSSR) technologies were found to encourage innovative consumers in Generations X and Y. However, it was emphasized that Generation Y is more innovative than Generation X in adopting embedded AI-MSSR applications for fashion retail shopping (Arachchi & Samarasinghe, 2024).

Personal income level also affects expectations from mobile shopping and consumption behavior. High-income consumers tend to favor premium mobile applications that offer personalized services and exclusive collections. Budget-conscious consumers are drawn to discounts, promotions, and fast-fashion brands (Azemi et al., 2022).

The study conducted in Spain revealed a positive correlation between daily internet usage that was used as an indicator of education level, digitalization, and digital literacy—and the penetration rate of online fashion buyers, reflecting the growth of fashion e-commerce and online purchasing (Martin, 2024). On the contrary, the findings of a study conducted to empirically determine the key factors influencing

US consumers' intentions to use apparel mobile applications revealed that there is a negative relationship between education level and the use of mobile applications for apparel purchases. This is mainly attributed to the fact that consumers with higher levels of education are less likely to trust apparel m-commerce. Consumers with higher education levels are more concerned about security and privacy issues when using apparel mobile applications in the US online apparel retail market (Sun & Chi, 2018).

In addition to different demographic profiles, fashion consciousness has an influence on the use of mobile applications for apparel purchasing. Consumers who follow fashion trends and prioritize style over functionality often rely on mobile shopping for quick access to the latest collections. Frequent clothing shoppers are more likely to use mobile applications for both browsing and purchasing, especially during sales or new product launches. The convenience and personalized experiences offered by mobile applications cater to the preferences of these consumers. These consumers are also drawn to features like virtual fitting rooms and personalized style recommendations. Virtual fitting rooms, utilizing AI and AR technologies, allow customers to try on clothes virtually, enhancing their shopping experience and reducing returns (Sapra, 2024).

4.2 *Consumer Motivations in the Use of Mobile Shopping Applications*

Mobile devices such as cell phones have their constraints such as small screens, inconvenient input, and limited computing abilities, which make it difficult for mobile users to search for and view relevant information. Besides, mobile shopping applications have greater risk as they are more vulnerable to a hacker attack and information interception and the encryption systems are not as intact and robust as online encryption systems, which increases consumers' concern about protecting personal information, location privacy, and payment security. On the other hand, the low switching cost for consumers makes it difficult for retailers to retain their consumer base (Gao et al., 2015).

Despite these risks, statistics revealed the rise of a digital population. Almost 59.5% of the global population are active internet users and 92.6% of them are considered as active mobile application users (Pop et al., 2023). The increase in time spent on mobile devices contributed to the growth of m-commerce and mobile shopping applications, especially the applications for apparel products in comparison to other generic mobile applications (Soni et al., 2019). On the other hand, the apparel industry is highly volatile and the companies must quickly respond to consumer demand by better understanding them (Koch et al., 2020). However, the time spent on using mobile shopping applications is not the direct measure for these devices to be used for making actual purchases as they can be seen as a kind of leisure activity. Therefore, the identification of underlying motivations for the adoption and future use of mobile shopping applications and providing more applicable findings are important. Besides, not only the intention to use and future use is important but also the intention to use them for actual apparel purchase is a critical issue to be investigated.

Consumers' motivations for using mobile shopping applications can be grouped as utilitarian and hedonic motivations and social influence, as in the features of online and mobile shopping applications.

Utilitarian motivations are practical, goal-oriented reasons for shopping online. These motivations include low prices and cost-saving, improved usability of product information, right assortment, and convenience. Online platforms often provide discounts, deals, and competitive pricing compared to physical stores. Detailed descriptions, reviews, and ratings help consumers make informed decisions. A wide variety of products, often categorized and easily searchable, enables consumers to find exactly what they need. Online shopping eliminates the need for physical travel, providing access to stores anytime from any location (Dorie & Loranger, 2020; Agrawal, 2022). With regard to mobile fashion retail applications, the dimensions of utilitarian value motivations include "efficiency shopping," and "achievement shopping." These dimensions cover time savings or the reduction of other relevant resources during a shopping, goal-oriented shopping for a particular purpose, and cost/value aspects (Parker & Wang, 2016). The findings of a research that identified the factors that impact customer experience while using fashion retailers' m-commerce applications emphasized the significance of utilitarian aspects of technology, including user-friendliness, convenience, and the ability to personalize the experience (McLean et al., 2018).

Hedonic motivations are experiential and pleasure-oriented reasons for shopping online such as fun, entertainment, enjoyment, and playfulness. The dimensions of hedonic value motivations for mobile fashion retail applications are defined as "gratification shopping" (seeking for good feeling or specific treat), "social shopping" (impact of important individuals and social relationships), "discount shopping" (shopping for bargain hunting or discount seeking), "idea shopping" (information collection of new trends and fashion), "role shopping" (selecting right products for others), and "adventure shopping" (try new things and get new ideas) (Parker & Wang, 2016).

Many consumers find browsing and exploring online platforms enjoyable. The interactive and engaging elements of websites and mobile shopping applications, such as gamification, personalized recommendations, and social sharing, make shopping a pleasurable activity (Loupiac & Goudey, 2020; Agrawal, 2022). Studies contain findings that hedonic motivations significantly affect mobile application usage. People who are highly motivated by hedonic factors are more likely to use technology. Consumers link their shopping experience from mobile applications with fun, escapism, enjoyment, playfulness, and an overall sense of pleasure. They show a greater propensity to turn to an interactive, multimedia-rich, enjoyable, and sophisticated applications for their m-commerce requirements (Vinerean et al., 2022). When consumers discover that purchasing using mobile applications offers them a wide range of goods and services together with enjoyment and convenience, they begin to trust and embrace the process (Baabdullah, 2018). Hedonic motivation also has a substantial positive impact on application users' impulsiveness because it is linked to the joy, fun, and delight that come from buying activities (Chopdar et al., 2022). On the other hand, hedonic consumers have atmospheric needs and are motivationally oriented towards purchasing according to their perception of environmental cues (Lee & Kim, 2019).

The findings of a study, which examined the hedonic and utilitarian motivations for m-commerce fashion retail applications engagement demonstrated that utilitarian motivations are followed by hedonic motivations for customer preferences and engagement. When it comes to utilitarian motivations, "efficiency shopping" plays a crucial role in shaping customers' behavior, whereas "gratification" and "social shopping" were the most important determinants when using m-commerce fashion retail applications (Parker & Wang, 2016). Nevertheless, fast-fashion consumer attitudes are more strongly influenced by hedonic motivations compared to utilitarian motivations when shopping through fast fashion mobile applications (Pop et al., 2023). In addition, a recent study revealed that hedonistic purchasing orientation, customer need for mobile application atmospheres, and intention to reuse mobile applications are all positively correlated. Atmospherics in mobile applications have a mediating role in the relationship between hedonic shopping orientation and the intention to reuse mobile applications. Therefore, a mobile shopping application that is dynamic and interactive without causing irritation is a crucial tool for marketers looking to influence consumers' attitudes (Erdem et al., 2024).

Changing personality trait patterns across generational cohorts and their effects are also reflected in mobile shopping application usage motivations. Although some sources state that Generation Y prefers online shopping for hedonic reasons (Brown & Lubelczyk, 2018), they, especially women, perceive fashion mobile applications as utilitarian tools, rather than sources of hedonic experiences. For Generation Y women, conveniences specifically timesaving, access to information for easily finding items and discounts are the key drivers of fashion application adoption (Parker & Wenyu, 2019; Parker & Kuo, 2022). According to the results of a study with participants who were primarily from Generation Y, consumers' attitudes toward the use of fast fashion mobile applications are positively influenced by perceived usefulness and low privacy and security concerns. Customers are more inclined to purchase fast fashion items via a mobile application since it makes it possible for them to complete activities faster, improves their efficiency, and boosts their productivity. As a result, these utilitarian features enhance users' attitudes while they are using the applications (Pop et al., 2023). On the other hand, the majority of Generation Z has adopted a hedonic lifestyle and makes impulsive and irrational fashion purchases from e-commerce applications (Utomo et al., 2023). As for repeat purchase intentions through online channels to purchase clothing and beauty products, Generation X is more likely to make repeat purchases driven by utilitarian values, while Generation Z tends to make repeat purchases based on hedonic values (De Lima & Monteiro, 2022).

Social influence is another motivation affecting customers' tendencies and plans to use m-commerce. The degree to which a user believes that significant people, such as friends and family, think they should use and adopt various information systems, such as m-commerce is known as social influence (Chong, 2013). When influential people endorse and recommend m-shopping, customers are more likely to embrace and incorporate new technologies into their lives. Consumers' intentions to accept and use m-shopping are heavily influenced by the standards and beliefs of their peers within the community (Vinerean et al., 2022).

Generation Y consumers exhibit distinct behaviors in online apparel shopping, characterized by "fashion-hunting," "status and recognition seeking," "variety-seeking," and "deal-hunting" (Agrawal, 2023). Their purchasing decisions are driven by sociopsychological benefits and sensory experiences, with online research and content shared by their close social circles playing a significant role. As digital natives, Generation Y is highly influenced by social norms, particularly within peer groups in technologically advanced societies (Wei et al., 2021). While social media exposure impacts their awareness of fashion products, peer-generated electronic word-of-mouth (eWOM) does not necessarily dictate their purchases due to individual differences in personality and fashion preferences. However, they do report being influenced by fashion advertisements on social media, leading to actual purchases (Parker & Kuo, 2022). Furthermore, social influence significantly shapes the purchasing intentions of working women when it comes to fashion items marketed through social media (Sharma & Chaudhry, 2024).

Beyond shopping behavior, Generation Y demonstrates a strong motivation for brand affiliation, preferring to engage with brands that align with their lifestyle and identity (Bento et al., 2018). Having grown up in an interconnected, technology-driven society, they are highly engaged with digital communication and consume more social media brand content than Generation X (Bento et al., 2018; Ladhari et al., 2019). They actively seek information through digital platforms and frequently interact with brands and retailers on social media (Bolton et al., 2013). Additionally, they often use these platforms to gain peer approval, reinforcing their engagement with online brand interactions. Overall, Generation Y's fashion shopping behavior is shaped by a combination of digital engagement, social influence, and personal preferences, making them a key segment for online fashion retailers.

The motivations driving the intention to purchase fashion products via fashion mobile applications vary based on cultural differences. Comparing the two samples, the influence of utilitarian factors on satisfaction is stronger among Italian users than Chinese users, suggesting that they tend to be more satisfied with their m-commerce experience if they can accomplish tasks easily, conveniently, and with personalization features fitting their needs. Chinese consumers, on the other hand, are more satisfied with the pleasure of searching fashion items through mobile applications. Moreover, Italian consumers, who belong to a culture characterized by a high level of uncertainty avoidance, are likely to spread eWOM if the m-commerce experience meets utilitarian motivations. On the contrary, the effect of utilitarian factors on eWOM was not significant in the Chinese sample (Hu et al., 2023).

The findings of a study conducted for analyzing consumers' willingness to use fashion mobile applications with regard to social influence showed that South Korean respondents are more influenced than the American respondents by family members and friends in the use of fashion mobile applications (Moon & Domina, 2015).

4.3 Adoption of Mobile Shopping Applications after COVID-19

The adoption of mobile shopping applications in the fashion industry surged significantly after COVID-19 due to the concerns over health risks, changing consumer attitudes and behavior, technological advancements, ease of digital access, and shifting market dynamics. The pandemic restricted physical store access, such as store closures, quarantine measures, and efforts to avoid crowded places to reduce virus transmission, making mobile applications a vital channel for purchasing fashion items. During the COVID-19 pandemic, mobile application solutions played a crucial role in reducing the risk of cross-contamination caused by close contact. Concerns about in-store safety, including trying on clothes and the risk of infection, further discouraged offline shopping. Social distancing measures also contributed to the shift, as consumers prioritized maintaining a safe distance in public spaces. Additionally, emotional factors played a significant role, as anxiety and low moods caused by the outbreak led consumers to seek comfort through online shopping. Overall, these situational factors significantly influenced the preference for online fashion shopping during the pandemic (Sumarliah et al., 2022).

In a study conducted in Romania, respondents generally exhibited rational purchasing behavior regarding items like clothing, which were categorized as nonessential during the pandemic. As a result, 56.6% of those surveyed reported spending less than before the pandemic, 38.2% maintained the same spending level, and 5.3% indicated that their clothing expenses were higher compared to the pre-pandemic period. An increased sensitivity occurred toward discounts offered by retailers particularly among low to medium-income individuals. This shift was driven by tighter budgets and the ease of accessing online deals. Under these circumstances, the digital environment like mobile applications enhanced exposure to personalized offers through targeted advertisings, mobile alerts, and social media campaigns, further shaping purchasing decisions. During the pandemic, higher-income earners, while still considering price reductions, tend to prioritize quality and exclusivity in their shopping experience. These varying attitudes accelerated the development of diversified strategies for each consumer segment in digital shopping environments. The evolution of the fashion business, including sustainability elements, was of moderate to high interest to a considerable proportion of respondents (more than 50%) (Demyen, 2024). In essence, this result put forward that the pandemic has impacted consumer thinking about the fashion industry.

The factors behind channel switching intentions and actual behavior of consumers from offline to online were defined as perceived severity of COVID-19, altruistic fear, response efficacy, and self-efficacy. Besides, age played a moderating role, with younger consumers responding more favorably to online switching (Youn et al., 2021). A study conducted during the COVID-19 pandemic found that online shopping intention was positively influenced by attitudes toward e-shopping, social norms, perceived behavioral control, and fear of in-store shopping. However, for future purchasing behavior, only attitudes toward e-shopping and perceived behavioral control played a significant role (Theodorou et al., 2023).

There are also studies in the literature that reveal how the COVID-19 pandemic has affected hedonic, utilitarian and social influences on online fashion shopping. A study by Vinerean et al. (2022) in Romania found that performance expectancy, hedonic motivation, and social influences play a crucial role in shaping consumers' intention to use mobile commerce during COVID-19. The findings indicated that consumers' willingness to engage in mobile shopping is largely driven by their perception of the platform's effectiveness in enhancing productivity, streamlining shopping experiences, and facilitating timely purchases. Additionally, m-commerce applications offer users an enjoyable and entertaining shopping experience. Moreover, in uncertain and potentially risky health situations, individuals tend to rely on family and friends for information processing and decision-making, including the adoption of mobile commerce platforms. Similarly, the examination of the effects of mobile applications on consumer purchase intentions in the fast fashion industry revealed that hedonic and social motivations played a key role in mobile shopping adoption during COVID-19 pandemic, with users prioritizing enjoyment, convenience, and social influence over utilitarian benefits (efficiency, time-saving) (Pop et al., 2023). Brand trust and familiarity became crucial in application-based shopping (Gil-Cordero et al., 2023). In contrast, the analysis of the data gathered from Indonesian consumers revealed that both hedonic and utilitarian motivations significantly impacted consumers' intent to buy fashion products online, but the utilitarian motivation was a stronger predictor. This suggests that consumers preferred online channels for purchasing fashion products due to their convenience, efficiency, and usefulness during the COVID-19 pandemic (Sumarliah et al., 2022).

5 Empirical Study

5.1 Research Model and Data Analysis

In this empirical study, consumer behavior in the adoption and future use of mobile shopping applications dedicated to apparel products was analyzed. Specifically, the tendency for future use of these applications for actual purchase was investigated. Besides, the adoptions of mobile shopping applications because of fear of COVID-19 were investigated with a multigroup analysis applied to two groups of consumers affected from COVID-19 pandemic at different levels.

Numerous studies were established about mobile commerce and mobile shopping applications in literature. But the adoption of mobile commerce should be investigated as a product specific phenomenon (Singh & Srivastava, 2020), therefore, in this section only the studies established in apparel retailing were considered, which differed in terms of purpose and scope; as well as the model and variables employed by them.

The majority of the studies focused on measuring the impact of different motivations on adoption behavior (Lian & Yen, 2014; Koch et al., 2020; Wu & Song, 2021), which are followed by the studies investigating the continuance behavior of the consumers to use mobile shopping applications (Gao et al., 2015; Mahapatra, 2017; Wu & Song, 2021). However, none of these studies investigated the adoption and future use of mobile shopping applications for the actual purchase of apparel products.

Besides, the studies differed in terms of the characteristics of the participants. In this respect, some studies presented the findings gathered from participants living in different countries (Wu & Song, 2021; Reyes-Mercado et al., 2017; Milanova & Saprikis, 2018). Moreover, participants belonging to different generations were investigated and younger consumers' behaviors were compared with that of their older counterparts (Lian & Yen, 2014; Koch et al., 2020; Wu & Song, 2021). Furthermore, some studies were established during the COVID-19 pandemic, and they revealed the main mechanism of consumer adoption of mobile shopping applications in this specific period (Vinerean et al., 2022; Pop et al., 2023). Specifically, Vinerean et al. (2022) validated the applicability of the unified theory of acceptance and use of technology (UTAUT) model in examining consumers' behavioral intent of using m-commerce during a pandemic. Pop et al. (2023) employed the stimulus organism response algorithm (S-O-R) algorithm and technology acceptance model (TAM) as a framework and the authors analyzed the moderation of COVID-19 within the relationship between attitude and purchase intention. However, in none of these two studies, the response level of participants was compared according to their affection level from the COVID-19 pandemic to show if the impact of the COVID-19 pandemic is temporary.

The majority of the studies implemented TAM, theory of planned behavior (TPB), and UTAUT models. In this respect, the researchers employed individual models on a single base (Reyes-Mercado et al., 2017; Chi, 2018; Milanova & Saprikis, 2018; Wahyudin & Awalludin, 2020; Soni et al., 2019). Some researchers benefited from the integration of models (Sun & Chi, 2018; Sun & Chi, 2019). Specifically, two models TPB and TAM, which are derived from the theory of reasoned action (TRA) and capable of defining the main determinants of technology adoption, were combined together (Nigam & Yamini, 2022; Singh & Srivastava, 2020). TPB and TAM became the most cited theories in the majority of research works conducted in e-commerce and m-commerce domains as well as the ones established in apparel retailing (Sun & Chi, 2019). Implying that the behavioral intention (BI) of an individual is dependent upon his/her behavioral attitude (BA), subjective norm (SUBNRM), and perceived behavioral control (PBC) referring to people's both internal and external constraints, TPB concentrates on individuals thought and beliefs. Expressing perceived usefulness (PU) and perceived ease of use (PEOU) as the main determinants of an individual's intention to use a particular technology, TAM concentrates on the individual perceptions about the characteristics of technology. Bringing these two models together enhances the explanatory power of the two models (Wahyudin & Awalludin, 2020).

In addition to the use of different models, the researchers extended the models with variables selected for a particular purpose. The trust had the highest frequency being selected as an additional variable (Reyes-Mercado et al., 2017; Milanova & Saprikis, 2018; Singh & Srivastava, 2020; Sun & Chi, 2018; Gao et al., 2015).

Considering the points above, a conceptual model integrating two models of TPB and TAM might put forward both the thoughts and beliefs of consumers as well as the impact of characteristics of adopted technology. Employing the core factors of TAM (PEOU and PU) as the antecedents for the factors of TPB (BA, SUBNRM, and PBC) and finally BI for future use as the dependent variable might be built an algorithm as in the case of S-O-R facilitating the interpretations. Besides, the conceptual model having the factor actual behavior for purchasing apparel products (AB)" as the dependent variable and the factor trust (TRUST) as an antecedent for AB might enable us to understand the future use of mobile shopping applications for actual purchase after the effects of the pandemic period passed away. Furthermore, analyzing the model for two groups of consumers affected by the COVID-19 pandemic at different degrees enables us to understand, if the effect of this kind of event is temporary or not.

The COVID-19 pandemic had many effects on human psychology. The fear and anxiety regarding the health and economic conditions interrupted social life. The studies established during the pandemic period implied that the COVID-19 pandemic has an impact on the enhancement of the use of e-commerce and mobile shopping applications (Koch et al., 2020; Vinerean et al., 2022; Pop et al., 2023; Ongkadinata et al., 2022). Considering the implications of the researchers, the following hypothesis was developed:

H1: The level of fear will influence all of the relationships in the adoption and use of mobile shopping applications

The ease of use and/or convenience of mobile shopping applications was presented to have a major role in the success of mobile shopping applications. Thus, PEOU is indicated to be a robust determinant to adopt mobile shopping applications and continue to use them (Nigam & Yamini, 2022). On the other hand, new technology providing more convenience and functionality in life becomes more attractive for consumers, who desire to explore new products to save time and improve efficiency in their personal life (Sun & Chi, 2019). Previous studies confirmed the positive influence of PU on consumers' positive attitudes (Pop et al., 2023). Therefore, hypotheses H2 and H3 were proposed:

H2: Perceived ease of use has a positive impact on behavioral attitude
H3: Perceived usefulness has a positive impact on behavioral attitude

Individuals have a higher likeliness to achieve certain behaviors when their attitude is positive towards a behavioral response. Specifically, the studies about mobile shopping applications for apparel products confirmed that users' positive attitudes initiated the usage and purchase intention (Chi, 2018). In addition, consumers follow the shopping trends and shopping experiences of significant people around them such as friends and family members (Ongkadinata et al., 2022). Many studies have confirmed the positive impact of subjective norms on the user's intention to adopt and utilize mobile shopping applications (Nigam & Yamini, 2022; Sun & Chi, 2019; Singh & Srivastava, 2020). However, people's confidence in their ability to perform activities strongly influences their behavior (Wu & Song, 2021). Thus, hypotheses H4, H5 and H6 were proposed:

H4: Behavioral attitude has a positive impact on behavioral intention
H5: Subjective norm has a positive impact on behavioral intention
H6: Perceived behavioral control has a positive impact on behavioral intention

Behavioral intention does not mean actual behavior (McLean et al., 2020). The causal link between intention and actual behavior was frequently disregarded in past studies (Singh & Srivastava, 2020). However, a customer's behavior can often be predicted by his or her intentions based on TRA (McLean et al., 2020). Thus, hypothesis H7 was proposed:

H7: Behavioral intention has a positive impact on actual behavior for purchasing apparel products

The impact of trust on intention was investigated in both the adoption phase and the continuance of behavior (Singh & Srivastava, 2020; Vinerean et al., 2022). Experienced consumers tend to display a higher willingness to continue to use mobile shopping applications as trustworthiness shapes users' mental expectations to think that m-commerce can deliver dependable service. Therefore, the user's intention to continue using m-shopping is impacted by trust (Vinerean et al., 2022). Thus, hypothesis H8 was proposed:

H8: Trust has a positive impact on actual behavior for purchasing apparel products

The definitions of the variables in the conceptual model are as below:

- **Perceived Ease of Use (PEOU):** It refers to the consumer's perception that mobile shopping applications do not require a lot of effort.
- **Perceived Usefulness (PU):** It refers to an individual's confidence that utilizing a mobile shopping application will enhance their performance in this study.
- **Behavioral Attitude (BA):** It refers to a complete evaluation indicating whether or not an individual likes to use mobile shopping applications.
- **Subjective Norm (SN):** It refers to people's perceptions in making decisions in terms of agreement or disagreement about using mobile shopping applications based on perceptions of other people such as friends, family members, coworkers, religious organizations, and others.
- **Perceived Behavioral Control (PBC):** It refers to the image of a person's both internal and external constraints on behavior, which is reflected in the individual's intention to use mobile shopping applications.
- **Behavioral Intention (BI):** It refers to the degree of readiness held by a consumer to continue to use mobile shopping applications.
- **Trust (TRUST):** It refers to the degree of confidence a consumer has in mobile shopping applications regarding safety and privacy concerns.

- **Actual Behavior (AB):** It refers to the actual use of mobile shopping applications for purchasing apparel products after the COVID-19 pandemic.

The data for the study was obtained via a survey conducted online in Turkey. The data was collected in a three-week period of time with a target of reaching at least 500 participants. The convenience sampling method along with snowball sampling was used in the study. Only the participants having made purchases from mobile shopping applications were considered.

A structured questionnaire having two parts was designed for the study. The questions about the socio-demographic characteristics and the approach of the participants towards COVID-19 pandemic took place in the first part. The fear of the COVID-19 pandemic was measured by asking questions based on the previous studies in the literature (Jian et al., 2020; Tran, 2021; Severo et al., 2021). The questions for analyzing the adoption behavior took place in the second part. Specifically, 4 questions for PEOU (Chi, 2018), 4 questions for PU (Chi, 2018), 3 questions for SUBNRM (Koch et al., 2020), 5 questions for BA (Han et al., 2010; Hansen, 2008), 4 questions for PBC (Crespo & del Bosque, 2008; Han et al., 2010), 4 questions for BI (Chi, 2018; Akar & Dalgic, 2018), 3 questions for TRUST (Troise et al., 2021) were prepared based on literature. 4 questions for AB were modified from the related literature (Akar & Dalgic, 2018). The questions about social demographic characteristics were prepared with convenient scales whereas the rest of the questions were established using a five-point Likert scale.

Data-set was analyzed using IBM SPSS 26 and IBM AMOS 22 statistical programs. The factors in the measurement model were extracted in exploratory factor analysis (EFA) and then confirmed by using confirmatory factor analyses (CFA).

The univariate normality was tested with the calculation of the skewness and kurtosis values for the data and the multivariate normality was examined with the Mardia's test. The scale variability and sampling adequacy were assessed by using Cronbach's alpha (CA), Kaiser-Myer-Olkin (KMO) measure, and Bartlett's test of sphericity. The confirmation of the factors was made with the evaluation of the goodness of fit values in CFA. The measurement model was checked to fulfill internal consistency, convergent validity, and divergent validity. Structural equation modeling (SEM) was used to test the hypotheses and find out the strength of the relationships. The predictive accuracy of the model was assessed with the coefficient of determination.

The participants were grouped based on differences in their fear of COVID-19. Toward this aim, the mean values of the responses calculated for the eight questions about fear of COVID-19 were calculated. A hierarchical cluster analysis followed by K-means cluster analysis was employed in the determination of the groups. A multigroup invariance test was employed to assess the equivalence in the measurement model between the groups, and the responses of the groups were compared.

5.2 *Results and Discussion*

The data was collected from 509 participants, whose profiles were given in Table 2.

Univariate normality was ensured for the data set as all the data had skewness values <3 and kurtosis values <10. CA was found to be 0.966 for the data set, which was well above the acceptable value of 0.7. The sampling adequacy was sufficiently large since KMO was measured as 0.948. Besides, the results of Bartlett's test of sphericity were found to be significant for the data set (15899.40 d.f.435, 0.001). The EFA was established by using the principal axis factoring with varimax rotation and 1 question from the factor PU was dropped because of having factor loading lower than 0.4. The rest 30 questions yielded seven factors explaining 75.079% of the total variance.

The multivariate normality was assumed as the multivariate kurtosis got the value of 620.695 for 31 questions, which was lower than the calculated value of 960 (30*32) based on the formula p(p + 2), where p equals the number of questions in the measurement model. The CFA based on Maximum Likelihood Estimation was applied and the goodness of fit values in the measurement model were calculated. Modification indices for error terms were checked and the error terms having high shared variance were correlated to improve Model Fit. The modified model yielded better model fit values (CMIN: 920.238; df: 365; CMIN/df:2.521 < 3 as recommended by Kline (2005); GFI:0.894 > 0.9 as recommended by Hair (1998); AGFI:0.865 > 0.8 as recommended by Chau (1996); CFI:0.965 > 0.9 as recommended by Hair (1998); RMSEA:0.055 < 0.08 as recommended by Browne and Cudeck (1992).

The results of convergent and discriminant validity analyses were summarized in Table 3.

The internal consistency was assumed since Cronbach alpha (CA) values of all factors ranged between 0.755 and 0.962. The convergent validity was confirmed as composite reliability (CR) values of all factors had a higher value than 0.6, and average variance extracted (AVE) values of all factors were within 0.5. The discriminant validity was also satisfied for all factors as the square root of AVE values was lower than the correlation between the factor and the other factors.

The cluster analysis yielded two clusters for the level of fear. The distance between initial clusters of 1.78. One-way ANOVA confirmed significant differences between the two clusters based on fear of COVID-19 (F = 1027.165, p < 0.000). The mean, standard deviation, and sample size were (4.40; 0.468; 407) for Cluster 1 corresponding to a high level of fear, and (2.62; 0.61; 102) for Cluster 2 corresponding to a low level of fear. The profile of each cluster is shown in Table 2. For the multi-group analysis, the metric invariance test as suggested by Kline (2005) was conducted and it resulted in a significant p-value (0.033) as seen in Table 4. The test implied that the groups were different at the model level and path differences should be checked. This indicated that the level of fear influenced the relationships in the adoption and use of mobile shopping applications for apparel products (H1 was supported).

Table 2 Sociodemographic profile of all participants, Cluster 1 and Cluster 2

		All participants		Cluster 1		Cluster 2	
		Frequency	%	Frequency	%	Frequency	%
Gender	Male	178	35	127	31.2	52	50
	Female	325	63.9	275	67.6	50	49
	Not answered	6	1.1	5	1.2	1	1
Age	18–24	159	31.3	127	31.2	33	31.4
	25–34	135	26.5	108	26.5	27	26.5
	35–44	133	26.1	103	25.3	30	29.4
	45 and above	82	16.1	69	17	13	12.7
Highest education achievement	High school/college	138	27.1	104	25.6	34	33.3
	Graduate	297	58.4	237	58.2	60	58.8
	Post graduate	74	14.5	66	16.2	8	7.9
Income level	Low	105	20.6	81	19.9	24	23.5
	Medium	284	55.8	237	58.2	47	46.1
	High	120	23.6	89	21.9	31	30.4
Expenditure on apparel as a share of disposable income	0–9%	201	39.5	160	39.3	41	40.2
	10–19%	139	27.3	108	26.5	31	30.4
	20–29%	87	17.1	72	17.7	15	14.7
	30% and above	82	8.2	67	16.5	15	14.7

Table 3 The results of convergent and discriminant validity analysis

	CA	CR	AVE	PEOU	PU	BA	SUBNRM	PBC	BI	TRUST	AB
PEOU	0.927	0.929	0.767	**0.876**							
PU	0.755	0.757	0.510	0.468	**0.714**						
BA	0.957	0.957	0.815	0.596	0.363	**0.903**					
SUBNRM	0.933	0.933	0.823	0.574	0.398	0.835	**0.841**				
PBC	0.868	0.874	0.635	0.772	0.430	0.606	0.571	**0.797**			
BI	0.962	0.964	0.872	0.369	0.313	0.769	0.678	0.463	**0.934**		
TRUST	0.927	0.927	0.809	0.573	0.352	0.555	0.769	0.555	0.646	**0.899**	
AB	0.919	0.906	0.708	0.506	0.417	0.736	0.720	0.595	0.712	0.686	**0.841**

Table 4 The GoF values of the baseline and constrained models and results of invariance test

	CMIN	df	CMIN/df	CFI	NFI	RMSEA	AIC	Δdf	ΔCMIN	p-value
Baseline model	1471.697	706	2.085	0.950	0.909	0.046	1919.697			
Constrained model	1527.860	735	2.079	0.948	0.906	0.046	1917.860	29	56.163	0.002

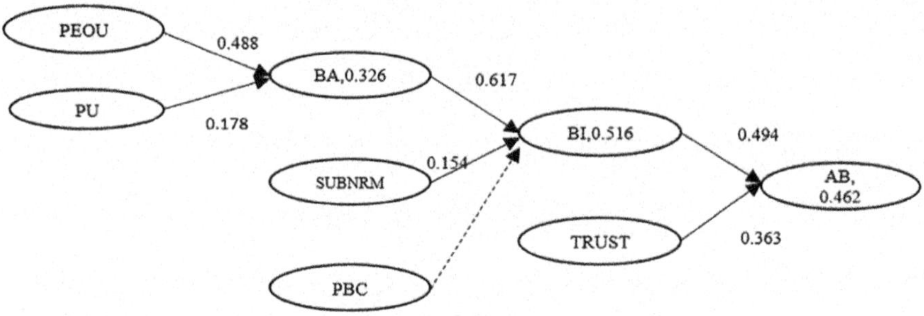

Fig. 2 Validated hypotheses and standardized regression weights for Cluster 1

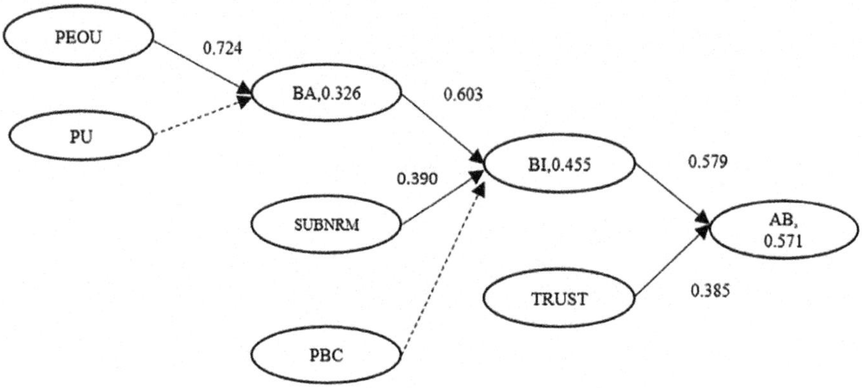

Fig. 3 Validated hypotheses and standardized regression weights for Cluster 2

The results of hypothesis testing, the estimates generated for each hypothesis, and R^2 value calculated for each factor were demonstrated in Fig. 2 for Cluster 1 (high fear level) and in Fig. 3 for Cluster 2 (low fear level), respectively.

For the participants included in Cluster 1, all of the hypotheses except H6 were confirmed at the 0.05 level of significance. PEOU was found to have a higher influence than PU on BA, which was more influential than SUBNRM on BI. Finally, BI had a higher influence than TRUST on AB. For the participants included in Cluster 2, all of the hypotheses except H3 and H6 were confirmed at the 0.05 level of significance. Unlike Cluster 1, the impact of PU on BA was found to be statistically insignificant. Moreover, the impact of SUBNRM on BI was much more apparent for Cluster 2. Similar tendencies have been observed within the relationships between BI, TRUST, and AB.

The present study revealed that PEOU was a significant factor influential on BA in line with the previous studies (Koch et al., 2020, McLean et al., 2020). Besides, BA and SUBNRM were highly influential on BI, which is in accord with previous findings (Sun & Chi, 2018; Akar & Dalgic, 2018; McLean et al., 2020; Ongkadinata et al., 2022; Pop et al., 2023). PBC was not considered as a primary motivation for

BI in parallel with the previous studies (Crespo & del Bosque, 2008; Nigam & Yamini, 2022).

On the other hand, the findings revealed that the consumers having different levels of fear behaved differently. For both groups, PEOU was found to be influential on BA but this relation had higher strength for the group of individuals having a lower level of fear. More interesting than that, the impact of PU was not found to be a prominent factor for the group having a low level of fear despite the contradictory findings in previous studies (Chi, 2018; Koch et al., 2020). These two findings present evidence for the inference that the main reason for the individuals having a high level of fear (Cluster 1) for the adoption and use of mobile shopping applications is their functionality during the COVID-19 pandemic. In fact, the lower impact of SUBNRM on BI for the same cluster (Cluster1) presents another support for the withdrawal attitude adopted by these individuals. The individuals within the isolation felt obliged to rely on their own experiences.

Another important finding of the study was that the selection of mobile shopping applications for the actual purchase of apparel products was much more favored for the group of individuals that have low levels of fear (Cluster 2). Even though the impact of BI on AB was agreed upon by both groups as already confirmed in previous studies (Akar & Dalgic, 2018), the larger strength value of this linkage for less fearful individuals proves that the fearful individuals mostly adopt the mobile commerce because of the temporary feeling raised because of COVID-19 pandemic.

Apart from the impact of BI on AB, TRUST was favored in both groups of individuals revealing its influence on the continual use of mobile shopping applications even after the COVID-19 pandemic. This meant that trust has an influence on preference for future use in addition to being an initial barrier to preventing individuals from adopting (Nigam & Yamini, 2022). This result was expected to some extent as trust was not considered for only BI but also considered as an antecedent for actual purchase. The importance of topics such as the prevention of privacy/security on purchase intention and actual behavior was already revealed in a limited number of past studies (Pop et al., 2023; Vinerean et al., 2022; Singh & Srivastava, 2020).

5.3 Practical Implications and Recommendations

The study is the first one to propose and verify an integrated model developed based on TAM and TPB models within the framework of S-O-R, which was extended with the factor TRUST and the actual behavior standing for the use of mobile shopping applications for purchasing apparel products after COVID-19 pandemic.

This study contributed to the literature in four ways. The integration of TAM into TPB exhibited a higher explanatory power by benefiting the strength of both models and eliminating their lack. The model was organized in an S-O-R framework, where the stimuli corresponded to the core factors of TAM, the organism corresponded to the core factors of TPB and; BI and AB stood for the response variables. Whereas TAM allowed analysis of the contribution of the characteristics of technology on the adoption of mobile shopping applications, the TPB model allowed analysis of the thoughts and beliefs of consumers. The factor AB as a response variable measuring the intentions of the consumers to use the mobile applications for purchasing apparel products justified the use of mobile shopping applications for making real shopping.

Second, actual behavior in the model analyzed the preference of mobile shopping applications even after COVID-19. Therefore, together with the first contribution, the extended integrated model did not only reveal the adoption behavior but also the continuance behavior of the consumers by eliminating the lack of model in regards to addressing the time between adoption and actual behavior in the future.

Third, the factor TRUST was included in the model as an antecedent for the factor AB instead of measuring its impact in the initial adoption phase as was done in the majority of previous studies in literature enabled the model to express the impact of trust for future uses.

Finally, this study revealed strong evidence to support that the consumers affected by the COVID-19 pandemic at different levels behaved differently in both the adoption phase and future use of mobile shopping applications.

The knowledge gained from this study helps to better understand the adoption of mobile shopping applications during the COVID-19 pandemic and the tendency of consumers to use mobile shopping applications for making actual purchases of apparel products after the COVID-19 pandemic. The results of this study provide many implications for promoting mobile shopping applications for apparel products.

The findings indicated that the COVID-19 pandemic triggered the tendency of the individuals, who are affected by the COVID-19 pandemic at higher degree than that of the individuals, who keep calm and therefore make their decisions in a more objective and rational way. However, the situation is not the same for the future use of mobile shopping applications after the COVID-19 pandemic and/or the use of the mobile shopping applications for real purchasing of apparel products. Nonetheless, the retailers are sure to benefit from the increased awareness of the consumers about mobile shopping applications and apply more practical solutions to keep the interest of all consumer base.

Based on the finding of this study implying that PEOU was a significant factor, the retailers might be advised to put more emphasis on the user-friendliness of mobile shopping applications especially for the individuals, which adopt more objective and rational approach toward the adoption of a specific technology. Specifically, user-friendliness might be improved for the current content of the mobile applications with the use of filtering and refinement tools. Moreover, the companies might be advised to create solutions for solving the inherent problem of mobile shopping applications, to have small screens with additional zoom functions and presentation of better layout. User-friendliness might also be provided for the innovative contents such as the ones employing virtual reality and augmented reality, which are gaining popularity.

However, the usefulness of technology should be more highlighted for people having higher levels of fear and anxiety. In this respect, it might be suggested that promotional activities should be executed for the stressful people to make them evaluate the functionality of these applications for different purposes except their usage during the isolation period raised by the COVID-19 pandemic. Considering that SUBNRM was found to be an influential factor on BI for both clusters, companies might be suggested to provide persuasive messages on social media platforms such as Instagram or Facebook to reach a large audience.

As pointed out in the results of this study, trust was carried out to be an important factor in actual behavior. Therefore, retailers should put great effort into increasing their reliability and trustfulness to guarantee the continual use of these platforms irrespective of consumer type. Considering the rising risks of data security, retailers might be advised to continue to make investments in the privacy and security of their users.

Despite the theoretical and practical contributions of this study to the literature, there are some limitations to be mentioned, which can be considered as possible directions for future studies. This research provides the findings of study established in a particular population; therefore, it cannot be generalized to whole populations. Turkish people are known to well adopt mobile commerce, but not all consumers living in different countries might have that kind of tendency. This limitation can be overcome by the establishment of the study in different geographical locations.

In this study, the participants answered the questions for mobile shopping applications, they have already used and they are used to. To enhance the insight obtained from the study, the model can be extended with different factors in relation to trust. In this respect, the positive image of brands and retailers about reliability might also be considered. In this way, the process of building trust is better understood and it is found out if the trust is associated with not only the previous experiences but also with the brands' or retailers' image.

Last but not least, the model can also be applied to measure the moderating effect of demographic variables such as age and generation, which might cause different approaches to the adoption and continual use of mobile shopping applications.

Acknowledgement The authors would like to thank to Abdulmetin TIRPANCI and Mustafa ALTINTAS for their contribution in conducting the survey of empirical study as a part of their Senior Design Project.

References

Agrawal, D. K. (2023). Increasing digital dissemination and online apparel shopping behavior of Generation Y. *Journal of Fashion Marketing and Management: An International Journal, 28*(1), 28–44.

Agrawal, D. K. (2022). Determining behavioural differences of Y and Z generational cohorts in online shopping. *International Journal of Retail & Distribution Management, 50*(7), 880–895.

Agrawal, S. (2021). Integrating digital wallets: Advancements in contactless payment technologies. *International Journal of Intelligent Automation and Computing, 4*(8), 1–14.

Akar, E., & Dalgic, T. (2018). Understanding online consumers' purchase intentions: A contribution from social network theory. *Behaviour & Information Technology, 37*(5), 473–487.

Arachchi, H. D. M., & Samarasinghe, G. D. (2024). Impact of embedded AI mobile smart speech recognition on consumer attitudes towards AI and purchase intention across Generations X and Y. *European Journal of Management Studies, 29*(1), 3–29.

Arumugam, S., & Wing, J. (2020). Antecedents of M-commerce user experience from a millennials perspective in a south African fashion retail context [Paper presentation]. In *Conference on Information Communications Technology and Society (ICTAS)* (pp. 1–6).

Asia Fashion Ecommerce Market Size and Share Analysis., https://www.coherentmi.com/industry-reports/asia-fashion-ecommerce-market/market-size. (accessed 21 Nov 2024).

Azemi, Y., Ozuem, W., Wiid, R., & Hobson, A. (2022). Luxury fashion brand customers' perceptions of mobile marketing: Evidence of multiple communications and marketing channels. *Journal of Retailing and Consumer Services, 66*, 102944.

Baabdullah, A. M. (2018). Consumer adoption of Mobile Social Network Games (M-SNGs) in Saudi Arabia: The role of social influence, hedonic motivation and trust. *Technology in Society, 53*, 91–102.

Babatunde, S. O., Odejide, O. A., Edunjobi, T. E., & Ogundipe, D. O. (2024). The role of AI in marketing personalization: A theoretical exploration of consumer engagement strategies. *International Journal of Management & Entrepreneurship Research, 6*(3), 936–949.

Bakewell, C., Mitchell, V. W., & Rothwell, M. (2006). UK Generation Y male fashion consciousness. *Journal of Fashion Marketing and Management: An International Journal, 10*(2), 169–180.

Barseghian, A. (2022). *How nike is using analytics to personalize their customer experience.* https://www.forbes.com/councils/forbestechcouncil/2019/10/07/how-nike-is-using-analytics-to-personalize-their-customer-experience/ (accessed 22 Jan 2025).

Behrooz, S. (2023). *Embracing the future: How luxury fashion brands are leveraging ai.* https://www.linkedin.com/pulse/embracing-future-how-luxury-fashion-brands-leveraging-sanaz-behrooz/ (accessed 22 Jan 2025).

Bento, M., Martinez, L. M., & Martinez, L. F. (2018). Brand engagement and search for brands on social media: Comparing Generations X and Y in Portugal. *Journal of Retailing and Consumer Services, 43*, 234–241.

Bolton, R. N., Parasuraman, A., Hoefnagels, A., Migchels, N., Kabadayi, S., Gruber, T., et al. (2013). Understanding Generation Y and their use of social media: A review and research agenda. *Journal of Service Management, 24*(3), 245–267.

Brown, M., & Lubelczyk, M. (2018). *The future of shopping centers.* AT Kearney. https://www.kearney.com/industry/consumer-retail/article/-/insights/the-future-of-shopping-centers-article (accessed 13 Dec 2024).

Browne, M. W., & Cudeck, R. (1992). Alternative ways of assessing model fit. *Sociological Methods & Research, 21*(2), 230–258.

Chaffey, D. (2014). *Boo.com case study—A classic example of failed ecommerce strategy.* https://www.smartinsights.com/digital-marketing-strategy/online-marketing-mix/boo-com-case-study-a-classic-example-of-failed-ebusiness-strategy/ (accessed 13 Dec 2024).

Chang, S. H., Chih, W. H., Liou, D. K., & Yang, Y. T. (2016). The mediation of cognitive attitude for online shopping. *Information Technology & People, 29*(3), 618–646.

Chau, P. Y. (1996). An empirical assessment of a modified technology acceptance model. *Journal of Management Information Systems, 13*(2), 185–204.

Chi, T. (2018). Understanding Chinese consumer adoption of apparel mobile commerce: An extended TAM approach. *Journal of Retailing and Consumer Services, 44*, 274–284.

Chong, A. Y. L. (2013). Predicting m-commerce adoption determinants: A neural network approach. *Expert Systems with Applications, 40*(2), 523–530.

Chopdar, P. K., Paul, J., Korfiatis, N., & Lytras, M. D. (2022). Examining the role of consumer impulsiveness in multiple app usage behavior among mobile shoppers. *Journal of Business Research, 140*, 657–669.

Crespo, A. H., & del Bosque, I. R. (2008). The effect of innovativeness on the adoption of B2C e-commerce: A model based on the theory of planned behaviour. *Computers in Human Behavior, 24*(6), 2830–2847.

de Lima, A. P., & Monteiro, N. A. (2022). E-commerce: The influence of hedonic and utilitarian motivations on generations X, Y, and Z. In *Impact of digital transformation on the development of new business models and consumer experience* (pp. 129–148). IGI Global Scientific Publishing.

Dean, E. (2024). *100+ social commerce statistics for 2025*, https://whop.com/blog/social-commerce-statistics/ (accessed 22 Feb 2025).

Demyen, S. (2024). The online shopping experience during the pandemic and after—A turning point for sustainable fashion business management? *Journal of Theoretical and Applied Electronic Commerce Research, 19*(4), 3632–3658.

Dianat, I., Adeli, P., Jafarabadi, M. A., & Karimi, M. A. (2019). User-centred web design, usability and user satisfaction: The case of online banking websites in Iran. *Applied Ergonomics, 81*, 102892.

Doh, S. J., & Hwang, J. S. (2009). How consumers evaluate eWOM (electronic word-of-mouth) messages. *Cyberpsychology & Behavior, 12*(2), 193–197.

Dorie, A., & Loranger, D. (2020). The multi-generation generational differences in channel activity. *International Journal of Retail & Distribution Management, 48*(4), 395–416.

Fashion eCommerce in Asia: Top 5 Countries by Revenue https://ecommercedb.com/insights/fashion-ecommerce-in-asia-top-5-countries-by-revenue/4889 (acccessed 22 Oct 2024).

Erdem, Ş., Gülöz, G. E., & Çelik, Z. (2024). A study on consumers' intentions to reuse mobile apparel applications. *Fırat University, Journal of Social Sciences, 34*(2), 907–918.

Esmaeili, L., & Hashemi, G. S. A. (2019). A systematic review on social commerce. *Journal of Strategic Marketing, 27*(4), 317–355.

Europe Online Fashion (Apparel, Footwear) Market. (2024). https://www.globenewswire.com/news-release/2024/11/25/2986491/28124/en/Europe-Online-Fashion-Apparel-Footwear-Market-2024-A-198-77-Billion-Industry-by-2028-Fueled-by-Multi-Channel-Demands-Immersion-E-commerce-and-Artificial-Intelligence-Integration.html (accessed 24Apr 2025).

Fashion E-commerce Global Market Report. (2025), https://www.thebusinessresearchcompany.com/report/fashion-ecommerce-global-market-report (accessed 24 Apr 2025).

Feinberg, R., & Kadam, R. (2002). E-CRM web service attributes as determinants of customer satisfaction with retail web sites. *International Journal of Service Industry Management, 13*(5), 432–451.

Förnges, J., & Sieling, B. (2023). *Implementing CSR in fashion companies: Increasing green trust during the transformation process towards sustainable development: An exploratory analysis on the influencing factors of green trust in the fashion sector*.

Gao, L., Waechter, K. A., & Bai, X. (2015). Understanding consumers' continuance intention towards mobile purchase: A theoretical framework and empirical study–A case of China. *Computers in Human Behavior, 53*, 249–262.

Gil-Cordero, E., Ledesma-Chaves, P., & Baena-Luna, P. (2023). Acceptance factors of Zara's shopping app among fashion consumers during COVID-19. *Journal of Consumer Behaviour, 22*(4), 955–970.

Guo, J., Zhang, W., & Xia, T. (2023). Impact of shopping website design on customer satisfaction and loyalty: The mediating role of usability and the moderating role of trust. *Sustainability, 15*(8), 6347.

Hair, J. F. (1998). *Multivariate data analysis* (p. 664). Pearson.

Han, H., Hsu, L. T. J., & Sheu, C. (2010). Application of the theory of planned behavior to green hotel choice: Testing the effect of environmental friendly activities. *Tourism Management, 31*(3), 325–334.

Hansen, T. (2008). Consumer values, the theory of planned behaviour and online grocery shopping. *International Journal of Consumer Studies, 32*(2), 128–137.

Hinman, R. (2012). *The mobile frontier: A guide for designing mobile experience*. Rosenfeld Media.

https://ecommercegermany.com/blog/fashion-online-shopper-insights (accessed 17 Oct 2024a).

https://financesonline.com/amazon-statistics/ (accessed 22 Jan 2025a).

https://us.shein.com/bonus-point-program-a-371.html (accessed 22 Jan 2025b).

https://www.net-a-porter.com/en-tr/ (accessed 18 Oct 2024b).

https://www.paradergi.com.tr/dunya/2021/09/09/z-kusagi-10-yil-sonra-hepimizden-zengin-olacak/ (accessed 24 Nov 2024c).

https://www.retail-insight-network.com/news/puma-google-cloud-ai/ (accessed 22 Jan 2025c).

https://www.statista.com/forecasts/1344694/fashion-e-commerce-revenue-europe-country (accessed 22 Jan 2025d).

https://www.statista.com/statistics/1329338/clothing-shoppers-mobile-audience-united-states-age/ (accessed 03 Jan 2025e).

https://www.statista.com/topics/3481/fashion-e-commerce-in-the-united-states/ (accessed 15 Feb 2025f).

Hu, L., Filieri, R., Acikgoz, F., Zollo, L., & Rialti, R. (2023). The effect of utilitarian and hedonic motivations on mobile shopping outcomes. A cross-cultural analysis. *International Journal of Consumer Studies, 47*(2), 751–766.

Jian, Y., Yu, I. Y., Yang, M. X., & Zeng, K. J. (2020). The impacts of fear and uncertainty of COVID-19 on environmental concerns, brand trust, and behavioral intentions toward green hotels. *Sustainability, 12*(20), 8688.

Johnson, V. G. (2023). *Transitioning second-hand stores: The impact of technology and innovative application on sustainable fashion* (Master's thesis).

Kamaldeen, I. O. (2024). *The impact of content marketing on consumer's attitudes and purchase intention on fashion brands: A case study of Instagram and Tiktok video* (Doctoral dissertation).

Keru Consulting. (2023). *Revolutionizing retail: how ai empowers adidas in personalized email marketing*. https://www.linkedin.com/pulse/revolutionizing-retail-how-ai-empowers-adidas-personalized-w9prc/ (accessed 24 Apr 2025).

Kline, R. B. (2005). *Principles and practice of structural equation modeling*. Guilford Press.

Koch, J., Frommeyer, B., & Schewe, G. (2020). Online shopping motives during the COVID-19 pandemic—Lessons from the crisis. *Sustainability, 12*(24), 10247.

Ladhari, R., Gonthier, J., & Lajante, M. (2019). Generation Y and online fashion shopping: Orientations and profiles. *Journal of Retailing and Consumer Services, 48*, 113–121.

Lee, Y., & Kim, H. Y. (2019). Consumer need for mobile app atmospherics and its relationships to shopper responses. *Journal of Retailing and Consumer Services, 51*, 437–442.

Leung, K., & Antypas, J. (2001). Improving returns on m-commerce investments. *Journal of Business Strategy, 22*(5), 12–12.

Li, K., & Shen, F. (2023). Irritating or enjoyable? Exploring the effects of soft-text native advertising and social media engagement metrics. *Journal of Promotion Management, 29*(7), 992–1014.

Lian, J. W., & Yen, D. C. (2014). Online shopping drivers and barriers for older adults: Age and gender differences. *Computers in Human Behavior, 37*, 133–143.

Liang, T. P., Ho, Y. T., Li, Y. W., & Turban, E. (2011). What drives social commerce: The role of social support and relationship quality. *International Journal of Electronic Commerce, 16*(2), 69–90.

Lissitsa, S., & Kol, O. (2021). Four generational cohorts and hedonic m-shopping: Association between personality traits and purchase intention. *Electronic Commerce Research, 21*, 545–570.

Liu, C. Y., & Lai, J. Y. (2015). Improving customers satisfaction to the development of user interface design for mobile shopping through QFD. In *The asian conference on psychology and behavioral sciences 2015 official conference proceedings* (pp. 277–297).

Loupiac, P., & Goudey, A. (2020). How website browsing impacts expectations of store features. *International Journal of Retail & Distribution Management, 48*(1), 92–108.

Lovejoy, B. (2019). *Instagram launches in-app shopping from major brands, incl. Burberry, Dior, Nike, Prada.* https://9to5mac.com/2019/03/19/instagram-checkout/

Magrath, V., & McCormick, H. (2013). Marketing design elements of mobile fashion retail apps. *Journal of Fashion Marketing and Management: An International Journal, 17*(1), 115–134.

Mahapatra, S. (2017). Mobile shopping among young consumers: An empirical study in an emerging market. *International Journal of Retail & Distribution Management, 45*(9), 930–949.

Martin, S. B. (2024). *Demographic shifts and implications on the landscape of online consumer behaviour in Spain. The fashion industry case*. Master's thesis, Universidade NOVA de Lisboa (Portugal).

May, A. (2024). *How P. Louise nailed its TikTok live strategy and earned $2 million in 12 hours.* https://cosmeticsbusiness.com/how-p-louise-earned-two-million-on-tiktok-shop

McLean, G., & Wilson, A. (2019). Shopping in the digital world: Examining customer engagement through augmented reality mobile applications. *Computers in Human Behavior, 101*, 210–224.

Mclean, G., Al-Nabhani, K., & Wilson, A. (2018). Developing a mobile applications customer experience model (MACE)-implications for retailers. *Journal of Business Research, 85*, 325–336.

McLean, G., Osei-Frimpong, K., Al-Nabhani, K., & Marriott, H. (2020). Examining consumer attitudes towards retailers'm-commerce mobile applications–An initial adoption vs. continuous use perspective. *Journal of Business Research, 106*, 139–157.

Miladinovic, J., & Hong, X. (2016). *A study on factors affecting the behavioral intention to use mobile shopping fashion apps in Sweden*.

Milanova, N., & Saprikis, V. (2018). An empirical investigation of adopters' perceptions toward M-commerce: The case of Bulgarian University students. *Journal of Marketing and Consumer Behaviour in Emerging Markets, 1*(7), 66–83.

Moon, E., & Domina, T. (2015). Willingness to use fashion mobile applications to purchase fashion products: A comparison between the United States and South Korea. *Journal of Textile and Apparel, Technology and Management, 9*(3).

Murillo-Zegarra, M., Ruiz-Mafe, C., & Sanz-Blas, S. (2020). The effects of mobile advertising alerts and perceived value on continuance intention for branded mobile apps. *Sustainability, 12*(17), 6753.

Neilon, S. (2024). *Understanding the digital fashion retail revolution*. https://www.syntaxdata.com/research/understanding-the-digital-fashion-retail-revolution (accessed 22 Jan 2025).

Nigam, P., & Yamini, K. K. (2022). Exploring consumer adoption of M-commerce: A study based on extended theory of planned behaviour. *Journal of Positive School Psychology, 6*, 10704–10720.

Ongkadinata, D., Susanto, Y. B., & Juliadi, R. (2022). Enhancing consumers' online purchase intention of clothing products through technological improvement during COVID-19 pandemic. In *ICEBE 2021: Proceedings of the 4th international conference of economics, business, and entrepreneurship, ICEBE 2021, 7 October 2021, Lampung, Indonesia* (p. 344). European Alliance for Innovation.

Öztopçu, K. (2024). *Tüketici Eğiliminde Farklı Bakış: Z kuşağını anlamak*. https://www.businessweek.com.tr/haberler/z-kusagi-satin-alma-gucu-ve-harcanabilir-gelir-acisindan-giderek-daha-guclu-hale-geliyor-2347449 (accessed 12 Jan 2025).

Parker, C. J., & Henninger, C. E. (2018). Enabling sustainable behaviours through M-Commerce app design: Focus on the fast fashion industry. In *Eco-friendly and fair* (pp. 69–79). Routledge.

Parker, C. J., & Kuo, H. Y. (2022). What drives generation-y women to buy fashion items online? *Journal of Marketing Theory and Practice, 30*(3), 279–294.

Parker, C. J., & Wang, H. (2016). Examining hedonic and utilitarian motivations for m-commerce fashion retail app engagement. *Journal of Fashion Marketing and Management: An International Journal, 20*(4), 487–506.

Parker, C. J., & Wenyu, L. (2019). What influences Chinesefashion retail? Shopping motivations, demographics andspending. *Journal of Fashion Marketing and Management: An International Journal, 23*(2), 158–175.

Pop, R. A., Hlédik, E., & Dabija, D. C. (2023). Predicting consumers' purchase intention through fast fashion mobile apps: The mediating role of attitude and the moderating role of COVID-19. *Technological Forecasting and Social Change, 186*, 122111.

Poushneh, A., & Vasquez-Parraga, A. Z. (2017). Discernible impact of augmented reality on retail customer's experience, satisfaction and willingness to buy. *Journal of Retailing and Consumer Services, 34*, 229–234.

Preece, J., Rogers, Y., & Sharpe, H. (2015). *Interaction design: Beyond human-computer interaction* (4th ed.). Wiley.

Rai, J., & Yadav, R. K. (2022). Factors affecting the working women's online impulsive buying behaviour in Gorakhpur District of Eastern Uttar Pradesh. *Journal of Positive School Psychology*, 191–205.

Reyes-Mercado, P., Karthik, M., Mishra, R. K., & Rajagopal. (2017). Drivers of consumer attitudes towards online shopping in the Indian market: Analysis through an extended TAM model. *International Journal of Business Innovation and Research, 13*(3), 326–343.

Rodríguez-Torrico, P., San-Martín, S., & San José-Cabezudo, R. (2019). What drives M-shoppers to continue using mobile devices to buy? *Journal of Marketing Theory and Practice, 27*(1), 83–102.

Roll, M. (2017), *Net-A-Porter—Revolutionizing global fashion retail*. https://martinroll.com/resources/articles/branding/net-a-porter-revolutionizing-global-fashion-retail/

Roy Dholakia, R., & Zhao, M. (2009). Retail web site interactivity: How does it influence customer satisfaction and behavioral intentions? *International Journal of Retail & Distribution Management, 37*(10), 821–838.

Ruane, L., & Wallace, E. (2013). Generation Y females online: Insights from brand narratives. *Qualitative Market Research: An International Journal, 16*(3), 315–335.

Sairam, V. (2024). *Shaping the future of retail: Zara's AI innovations*. https://www.linkedin.com/pulse/shaping-future-retail-zaras-ai-innovations-vignesh-sairam-uw11c/ (accessed 08 Feb 2025).

Salas, M. (2024). *Tap to shop: How H&M revolutionised social commerce with shoppable posts*. https://www.riibon.com/post/tap-to-shop-how-h-m-revolutionized-social-commerce-with-shoppable-posts (accessed 08 Feb 2025).

Sapra, Y (2024). *AI in fashion and apparel: real-time virtual fitting rooms and personalized shopping*, https://www.hashstudioz.com/blog/ai-in-fashion-and-apparel-real-time-virtual-fitting-rooms-and-personalized-shopping/ (accessed 22 Jan 2025).

Schroeders, S., & Visser, M. (2019). Customer acquisition: Sales and pricing. In *Digital marketing fundamentals* (pp. 342–382). Routledge.

Severo, E. A., De Guimarães, J. C. F., & Dellarmelin, M. L. (2021). Impact of the COVID-19 pandemic on environmental awareness, sustainable consumption and social responsibility: Evidence from generations in Brazil and Portugal. *Journal of Cleaner Production, 286*, 124947.

Shao Yeh, Y., & Li, Y. M. (2009). Building trust in m-commerce: Contributions from quality and satisfaction. *Online Information Review, 33*(6), 1066–1086.

Sharma, P., & Chaudhry, M. (2024). Examining the factors affecting the Behavioural intention of working women for fashion apparels in case of social media marketing in circular economy. *Gurugram University Business Review (GUBR), 4*(2).

Singh, S., & Srivastava, S. (2020). Exploring the moderating role of product type in Indian m-commerce fashion and electronics market. *International Journal of Logistics Systems and Management, 36*(1), 61–91.

Solberg Söilen, K. (2024). Checkout and payment options. In *Digital marketing* (pp. 107–115). Springer.

Soni, M., Jain, K., & Kumar, B. (2019). Factors affecting the adoption of fashion mobile shopping applications. *Journal of Global Fashion Marketing, 10*(4), 358–376.

Sumarliah, E., Usmanova, K., Mousa, K., & Indriya, I. (2022). E-commerce in the fashion business: The roles of the COVID-19 situational factors, hedonic and utilitarian motives on consumers' intention to purchase online. *International Journal of Fashion Design, Technology and Education, 15*(2), 167–177.

Sumiyana, S., & Komariyah, E. F. (2024). Satisfaction and trust should first be nested in commitment before affecting loyalty in the use of mobile shopping applications. *SAGE Open, 14*(4), 21582440241288020.

Sun, J., & Chi, T. (2018). Key factors influencing the adoption of apparel mobile commerce: An empirical study of Chinese consumers. *The Journal of the Textile Institute, 109*(6), 785–797.

Sun, J., & Chi, T. (2019). Investigating the adoption of apparel m-commerce in the US market. *International Journal of Clothing Science and Technology, 31*(4), 544–563.

Theodorou, A., Hatzithomas, L., Fotiadis, T., Diamantidis, A., & Gasteratos, A. (2023). The impact of the COVID-19 pandemic on online consumer behavior: Applying the theory of planned behavior. *Sustainability, 15*(3), 2545.

ThredUp Resale Report. (2023). https://cf-assets-tup.thredup.com/resale_report/2023/thredUP_2023_Resale%20Report.pdf (accessed 08 Feb 2025).

Tran, L. T. T. (2021). Managing the effectiveness of e-commerce platforms in a pandemic. *Journal of Retailing and Consumer Services, 58*, 102287.

Troise, C., O'Driscoll, A., Tani, M., & Prisco, A. (2021). Online food delivery services and behavioural intention–a test of an integrated TAM and TPB framework. *British Food Journal, 123*(2), 664–683.

Tupikovskaja-Omovie, Z., Tyler, D. J., Chandrasekara, S., & Hayes, S. (2014). Segmenting the UK mobile fashion consumer. In *13th international conference on mobile business, ICMB 2014*.

UNCTAD Report. (2021). https://www.adhugger.net/2021/05/03/unctad-report-global-e-commerce-up-to-27-7trillion-covid-19-boosts-online-retail-sales/ (accessed 18 Dec 2024).

Utomo, S. B., Jamali, H., Arief, I., Saputra, M. N., & Priambodo, C. G. (2023). Analysis of the influence of hedonic digital lifestyle on consumptive e-shopping behavior of generation Z through E-commerce applications. *Jurnal Sistim Informasi dan Teknologi*, 85–91.

Verhoef, P. C. (2021). Omni-channel retailing: Some reflections. *Journal of Strategic Marketing, 29*(7), 608–616.

Vinerean, S., Budac, C., Baltador, L. A., & Dabija, D. C. (2022). Assessing the effects of the COVID-19 pandemic on M-commerce adoption: An adapted UTAUT2 approach. *Electronics, 11*(8), 1269.

Wahyudin, Y., & Awalludin, D. (2020). Differences in drives of online shopping intention between young and middle-aged adult through the application of UTAUT model and gender as a moderation variable. *Prosiding ICSMR, 1*(1), 21–36.

Wei, L., Hong, Y., & Zeng, X. (2024). Exploring the influence of content marketing strategies on the expansion of the fashion second-hand market: A theoretical prediction study. *Journal of Fashion Marketing and Management: An International Journal, 28*(5), 950–979.

Wei, M. F., Luh, Y. H., Huang, Y. H., & Chang, Y. C. (2021). Young generation's mobile payment adoption behavior: Analysis based on an extended UTAUT model. *Journal of Theoretical and Applied Electronic Commerce Research, 16*(4), 618–637.

Wu, J., & Song, S. (2021). Older adults' online shopping continuance intentions: Applying the technology acceptance model and the theory of planned behavior. *International Journal of Human–Computer Interaction, 37*(10), 938–948.

Yang, H. (2020). *Motivational drivers for mobile shopping adoption: An empirical study of the Chinese generation Y consumers*. University of Wales Trinity Saint David (United Kingdom).

Yang, S. C. (2012). Mobile applications and 4G wireless networks: A framework for analysis. *Campus-Wide Information Systems, 29*(5), 344–357.

Yarmosh, K. (2011). *App savvy* (2nd ed.). O'Reilly Media.

Yoon, D., Choi, S. M., & Sohn, D. (2008). Building customer relationships in an electronic age: The role of interactivity of E-commerce web sites. *Psychology & Marketing, 25*(7), 602–618.

You, Y., & He, Y. (2023). How brands drive electronic word-of-mouth in an interactive marketing environment: An overview and future research directions. In *The Palgrave handbook of interactive marketing* (pp. 149–179). Palgrave Macmillan.

Youn, S. Y., Lee, J. E., & Ha-Brookshire, J. (2021). Fashion consumers' channel switching behavior during the COVID-19: Protection motivation theory in the extended planned behavior framework. *Clothing and Textiles Research Journal, 39*(2), 139–156.

Zietek, N. (2016). *Influencer marketing: The characteristics and components of fashion influencer marketing*.

Zwass, V. (2003). Electronic commerce and organizational innovation: Aspects and opportunities. *International Journal of Electronic Commerce, 7*(3), 7–37.

The Role of Social Marketing on Promoting Sustainable Fashion Consumption

Keywords Social marketing · Sustainable fashion · Consumer behavior · Self-benefit · Other-benefit

1 Introduction

The global climate crisis has led to lifestyle changes, and individuals have begun to adopt more sustainable and responsible consumption patterns over the past few decades. Meanwhile, the COVID-19 pandemic accelerated the questioning of consumption habits and became the driving force of changing behaviors toward sustainable consumption. One of the most important reflections of this change has been seen in fashion consumption, and more attention has begun to be paid to overconsumption and waste in the fashion industry. Because, the fashion industry accounts for 10% of total global carbon emissions, consumes large amounts of water and energy, produces large amounts of waste and microplastics, and has the fourth largest environmental impact after food, housing, and transportation (https://earth.org/fast-fashions-detrimental-effect-on-the-environment/; https://www.eea.europa.eu/publications/environment-and-climate-pressures-from). Hence, the industry has undergone a radical transformation over time, with producers and consumers increasingly prioritizing sustainability. Traditional fashion methods have been evaluated in response to the growing public awareness of environmental, social, and economic issues of sustainability. The idea of sustainable fashion has gained popularity within this paradigm, providing an environmentally sound, socially acceptable, and economically viable substitute for the fast fashion business model. In this sense, green fashion, eco-fashion, and ethical fashion terms are also used interchangeably with sustainable fashion. Although the terms green fashion and eco-fashion seem to be limited to environmental issues, and the ethical fashion term is limited to fair trade principles, the sustainability framework should be considered as a whole and should include all dimensions of the sustainability concept.

C. Saricam, N. Okur, *Changing Textile and Apparel Consumption in Transformative Era of Sustainability and Digitalization*, Sustainable Textiles: Production, Processing, Manufacturing & Chemistry,
https://doi.org/10.1007/978-3-031-97419-9_4

As a matter of fact, for example, ethical issues include ethical concerns such as human rights and social justice, as well as environmental protection (Pérez et al., 2022). Therefore, sustainable fashion refers to compliance with all components of the sustainability concept (social, environmental, economic) from production to disposal, and includes sustainable business practices, consumption behaviors, and marketing activities.

A sustainable fashion business model incorporates innovative design and manufacturing processes, ethical labor practices, transparent and traceable supply chains, localized production, sustainable material selection, and a dedication to a circular economy. These strategies help the fashion industry limit its negative effects on the environment while also promoting a fairer and more morally upstanding global sector. Sustainable fashion creates a precedent for a future in fashion that is more responsible and environmentally mindful by constantly embracing these ideals (Aakko & Koskennurmi-Sivonen, 2013). Many brands now act responsibly by emphasizing sustainability and trying to reduce their environmental impact, as well as taking part in solving social issues. In the meantime, consumers' behaviors regarding sustainable fashion consumption have been put into focus.

Sustainable fashion consumption has three phases; purchasing, usage, and disposal of the fashion products (Iran, 2018). Consumers' attitudes and purchase intentions towards sustainable fashion products are affected by the factors that can be divided into two main groups: product-related and consumer-related factors. Product-related factors include product attributes, variety and availability, price, and value (Pura, 2005; Chi, 2015; Teng et al., 2018). Consumer-related factors mostly arise from consumers' pro-environmental attitudes (including their knowledge, awareness, and concern about the environment) and social concerns. Higher environmental knowledge of consumers is associated with a more positive attitude toward sustainable fashion consumption. This positive attitude therefore affects the sustainable fashion purchase intention. Consumer interest in sustainable fashion is growing as a result of knowledge about the damaging effects of the fashion industry on the environment as well as the pollution that results from the production of textiles. Thus, the findings of several researches have revealed that consumers' intentions to purchase sustainable fashion products are significantly influenced by their level of environmental knowledge (Ho et al., 2020). Another important factor that may develop pro-environmental behavior in sustainable consumption is environmental concern, which refers to the sensitivity of individuals toward the consequences of their behaviors on the environment (Pinho & Gomes, 2023). This factor can have an impact on product choices as well as consumer behavior during the usage and disposal stages. According to the previous researches, there is a close relationship between environmental concern and sustainable fashion consumption behavior (Cowan & Kinley, 2014; Mo et al., 2018; Yang et al., 2017; Heo & Muralidharan, 2019). Similarly, consumers with higher environmental awareness prefer sustainable, more durable, and second-hand fashion items (Seibel et al., 2021). In addition, socially concerned consumers' preferences could be more inclined to sustainable fashion. All these consumer-related factors are individual barriers to consumers in sustainable fashion consumption. Therefore, although the level of these factors in the consumer depends on the consumer herself/himself, providing detailed information to the consumer, especially on points such as social issues, can develop a supportive attitude and engagement towards sustainable consumption (Polonsky, 2011). While other barriers also exist, for example, in many studies, the factor of willingness to pay a premium price has been evaluated as a significant barrier to consumers' consumption of sustainable fashion (Khan et al., 2024), marketing activities can be an important driver of consumer preferences towards sustainable consumption. In other words, sustainable consumption can be facilitated by effective marketing of sustainable fashion or behavioral change can be created (Hill & Lee, 2012).

The concept of social marketing, which aims to change the behavior of the target audience for the welfare of both individuals and society, comes to the fore at this point. Changing human behavior positively is a long-term process that involves marketing methods to achieve social change. Although it mostly refers to nonprofit marketing targeting socially focused offerings through advertising and is used by governments, nonprofit social organizations, and public organizations, it can also be used as a marketing tool by the companies that carry out their activities with an awareness of ethical and social responsibility. It can be considered as a way to persuade consumers to inquire about consumption expenditures and encourage sustainable consumption, especially in today's world where excessive consumption is questioned. Fashion consumption is one of the most environmentally criticized areas, and it is thought that the concept of social marketing will find a wider application area to support sustainable fashion consumption within the scope of the sustainability activities of brands, especially after the COVID-19 pandemic. In this way, consumers' caring senses, including care for self, care for nature, and community, can be triggered, and they can be motivated for sustainable consumption (Sheth et al., 2011).

This chapter deals with sustainability in fashion, key features and importance of social marketing and the activities of fashion brands in this regard, sustainable consumption within the context of consumer identity, and empirical testing of a model for investigating the consumers' behavior towards sustainable fashion consumption including the constructs of social marketing appeals (fear, guilt, and shame), and self-benefit, and other-benefit appeals through a field study. The behavior of consumers toward sustainable consumption of fashion products has been evaluated from various aspects; however, the effect of social marketing appeals on this behavior has not been examined. Thus, along with comprehensive literature on the subject, studying the effect of interaction with the consumer through socially oriented marketing appeals on sustainable fashion consumption behavior will make a new contribution to the field of study. Overall, this chapter is expected to be beneficial for researchers and academics who study sustainable fashion consumption, and for social marketers in creating communication and engagement strategies for public and nonprofit organizations and sustainable brands to facilitate consumers' adaptation to sustainable fashion consumption.

2 Sustainability in Fashion

2.1 Sustainable Fashion Business

The concept of sustainability in fashion first began in the early 1990s, when "green" issues appeared in textile and fashion publications, during the period coinciding with the 1992 United Nations Conference on Environment and Development, known as the Rio Earth Summit (Charter et al., 2023). Realizing the environmental damage of overproduction and overconsumption, brands incorporated the concept of sustainability into their business models by creating collections primarily using environmentally friendly fibers or by evaluating the environmental burden of different fibers with life cycle assessment. However, sustainable fashion is not limited to the raw materials used. In addition, high water usage and pollution of chemical processes used in dyeing and finishing processes are harmful to the environment. Therefore, companies have begun to develop and use environmentally friendly production processes. Moreover, transparency and traceability in the highly complicated global supply chain of the industry have become the core priorities of the sustainable fashion business. However, the issue is not only the environmental impacts but also ethical practices are addressed by sustainable fashion. With the mass production of cheap items with a short lifespan, fast fashion has further increased the fashion industry's destructive impact on the environment and fair-trade concerns (Overcash, 2012). Sustainable fashion aims to reduce the environmental and social impact of the industry. This entails tackling problems including overuse of resources, contamination of the environment, and abusive labor practices. Sustainable fashion emphasizes longevity, ethical sourcing, and responsible manufacturing in contrast to fast fashion. Therefore, every stage of production, from design to distribution, needs to be evaluated and transformed to satisfy the requirements of sustainability.

Firstly, when sustainability in design is considered, sustainable design in fashion can be examined under the following headings (Nayak, 2022; McKinney et al., 2020; Wang et al., 2023; Enes & Saygili, 2023):

1. Using environmentally friendly fibers and recycled materials.
2. Minimizing waste in pattern-making.
3. Using digital technologies to simulate the design process without a need for physical prototypes.
4. Making modular garment designs to extend product life and adapt the product to different styles and uses.

Material choice is the foundation of sustainable fashion production process. Various fibers such as organic cotton and BCI-certified cotton fibers, hemp, flax, lotus fiber, wool, silk, mushroom and corn fiber (natural fibers), bamboo, lyocell, Ecovero™ (regenerated cellulosic fibers), coffee ground fiber, orange fiber (fibers extracted from waste), and recycled polyester and polyamide can be used as sustainable materials in fashion products. These materials are produced in a way that has the least negative effect on the environment, frequently by avoiding the use of dangerous chemicals and using less water. The objective is to develop a closed-loop system that takes into account a material's whole lifecycle, from sourcing to disposal (Hur & Cassidy, 2019).

Challenges in conventional cotton production, such as excessive and/or misuse of pesticides, herbicides, and chemical fertilizers, and excessive water consumption, have been criticized for some time. For this reason, organic and BCI cotton cultivation has emerged as a sustainable solution for cotton fiber, which is one of the main fibers used in fashion products due to its many desirable properties. Organic cotton is used in medical and hygiene products, apparel, and home furnishing. It is commonly used in the production of medical items, including bandages, surgical dressings, ear swabs, baby care products and diapers, napkins, and makeup removal pads because of its hypoallergenic nature and soft texture ensuring the well-being of the users (https://www.cognitivemarketresearch.com/organic-cotton-market-report; Nayak, 2022). It is also preferred in apparel products due to its environmentally friendly production, and according to Cognitive Market Research, the desire for chemical-free products makes the apparel category the fastest-growing category. Today, various global brands like Patagonia, Levi's, Nike, Gap, H&M, and C&A, as well as fashion designers like Mara Hoffman, Eileen Fisher, and Stella McCartney, use sustainable cotton in their collections (https://www.cognitivemarketresearch.com/organic-cotton-market-report).

Bast fibers such as flax and hemp are also increasingly used in sustainable apparel production. Compared to conventional cotton fiber cultivation, hemp is recognized as a more sustainable fiber due to its less water requirement and land for production. In addition, the higher amount of yield with less fertilizer and pesticide usage makes hemp fiber a promising alternative material for sustainable apparel products (Lawson et al., 2022). Similarly, flax fiber can be grown with less use of pesticides and fertilizer and very little irrigation or even with rainwater (Kozlowski & Muzyczek, 2023). About the fiber properties, flax fiber is characterized by its unique touch, comfort, antibacterial, antistatic, and UV protection properties (Behera 2007; Foksowicz-Flaczyk & Walentowska 2008; Jackowski et al., 2003), and hemp fiber is recognized by its inherent comfort related properties such as high moisture absorption, breathability as well as antibacterial, anti-static, and UV protection properties (Stanković et al., 2019; Yang et al., 2013). The mentioned bast fibers have been increasingly preferred in denim products and take place in the collections of many global brands such as Levi's, Diesel, Wrangler, and Mavi (https://www.mavi.com/mavi-hemp; https://www.levi.com/; https://www.wrangler.com/; https://global.diesel.com/).

The increasing demand for sustainable and environmentally friendly products made of natural and organic materials allowed the use of wool fiber to increase slightly. The use of wool in athleisure wear has also contributed to the growth of wool fiber use in apparel. According to Census Insights, the Asia-Pacific region is expected to have significant growth in the apparel market due to rising consumer awareness about the benefits of wool. Besides, the trend of customized and tailored designs in fashion and the increase in disposable income in countries like China and

India has contributed to the expansion of wool fiber in the apparel market. (https://www.linkedin.com/pulse/wool-fabric-apparel-market-research-report-unlocks-analysis-wnt4e/).

Another fiber group that is preferred and attracts attention in sustainable fashion products is regenerated cellulosic fibers such as lyocell, bamboo, and Ecovero™. Among the rayon fiber types, the third-generation regenerated cellulose fiber called lyocell stands out among the sustainable fibers due to the use of organic solvents (NMMO: N-methylmorpholine N-oxide) in the production process, the recycling of solvent and water, and the completion of the process in a short time and less energy consumption. The sentence has been revised: Its high moisture absorption capacity, breathability, high wet strength and dimensional stability, and exceptional hand and drapeability make it preferred in apparel products (Periyasamy & Militky, 2020). The production steps of regenerated bamboo fiber are largely similar to viscose fiber. Therefore, regenerated bamboo fiber is also called bamboo viscose. In this sense, although there are doubts in terms of sustainability due to the chemical processes used in the production phase, bamboo is defined as one of the sustainable fibers because, bamboo is the fastest growing plant, it is renewable and decomposable without pollution, and does not require chemicals such as pesticides and fertilizers or excessive amounts of water in its production). Products made of bamboo fiber are on the market due to its unique properties such as inherent antibacterial and biodegradable properties, high moisture absorption capacity, softness, brightness, and UV protection features. Bamboo fiber provides comfort in a variety of applications thanks to its high moisture absorption capacity, breathability, and quick drying behavior because of its distinct microstructure. Today, regenerated bamboo fiber is used in products such as underwear, sportswear, t-shirts, and socks. It is preferred in summer clothes, especially to protect pregnant women and children from UV rays. Its natural antibacterial properties make it suitable for use in hygienic products as it does not require chemical additives to achieve antibacterial properties and therefore does not cause skin allergies. In addition, regenerated bamboo fiber is used in household items such as towels, bathrobes, duvet covers, blankets, and curtains (Prakash, 2020).

Ecovero™ is a viscose fiber brand developed by Lenzing, known for its environmentally friendly production process for a cleaner and more sustainable environment. The less environmental impact of the fiber is due to the 50% fewer carbon emissions compared to the conventional viscose fiber as confirmed by the Higg MSI, and its biodegradability (https://www.ecovero.com/). Ecovero™ makes the fabric breathable as it regulates the absorption and release of moisture; therefore, it supports the heat flow that the body should have. It is lightweight and can absorb moisture four times its weight. Thanks to these features, it is preferred especially in summer clothes (https://nuvonda.ch/en/pages/ecovero).

Recycled polyester, which is an emerging sustainable material and widely used in the collections of almost all global brands, especially sportswear brands, has gained great importance recently. Recycling used PET bottles is a viable way to lessen the negative consequences of both their production and disposal. Recycling reduces greenhouse gas emissions in addition to saving energy and reducing raw material use (Majumdar et al., 2020). Nike's path to zero carbon and zero waste includes the use of recycled polyester in most of their products such as hoodies, pullovers, jackets, vests, pants, tights, t-shirts, shorts, bras, swimwear, and socks. Besides, Nike Flyknit is a specially developed lightweight fabric that produces on average 60% less waste than traditional shoe upper production. Each shoe made from Flyknit has 6–7 plastic bottles in its upper part (https://www.nike.com/sustainability/materials). Adidas, which produced its first shoe from recycled materials in 2015, announced that 96% of polyester used in its products is recycled polyester by the end of 2023 (https://news.adidas.com/). In 2023, Under Armour produced more than 300 products made of recycled polyester. Its patented Tech Tee and Armour Fleece fabrics feature 100% recycled polyester (https://about.underarmour.com/en/Purpose/Sustainability/environment.html).

In the design and production stages, minimizing the fabric waste within the production line contributes significantly to the sustainable fashion business. Today, around the world more than 100 to 150 billion apparel items are produced per year (https://fashionunited.com/global-fashion-industry-statistics). Considering that amount, it is possible to say that by reducing cutting and sewing waste, which is defined as a type of pre-consumer waste, by even 1%, a considerable amount of fabric can be saved from being waste to a great extent.

Modular garment design is a way of designing for longevity as an opportunity to reduce the carbon footprints of apparel items. In modular garment design, interconnected clothing parts can be added or removed to obtain several different product choices. Especially when all the parts are made of a single type of fiber and do not contain too many decorative accessories, sustainable products that are easy to recycle can be created (Gwilt & Pal, 2017). Modular garment design allows product reconfiguration and reinvention through the use of a flexible core with removable and adjustable parts that facilitate change over time, and this gives clothes a second life and extends their lifespan (Mansour, 2017).

In transforming sustainable designs into sustainable fashion products, innovative production techniques are embraced to reduce their environmental impact. Waterless dyeing techniques, for instance, drastically cut down on the amount of water used for dyeing and finishing processes. Furthermore, technological developments have made it possible for circular manufacturing, which reduces waste and maximizes the use of recycled or upcycled resources. These methods lower the use of resources and waste production, which improves the fashion products' overall sustainability (Henninger et al., 2016).

Another key component in the production of sustainable fashion items is transparency. Companies that place sustainability at the center of their strategies work hard to give customers access to information about their supply chains. Because of this transparency, ethical and environmental standards are upheld at every stage of the production process. Certain manufacturers employ blockchain technology to facilitate traceability, thereby enabling customers to track an item from its point of origin to the store shelf (Henninger et al., 2016).

Unfortunately, production does not become sustainable just by reducing its environmental burden. The UN 2030 Agenda for Sustainable Development has accelerated the interest of the fashion industry in ethical concerns including human rights, social justice, and fair salaries. Today in the sustainable fashion business, providing safe and sweatshop-free labor conditions, paying fair salaries, and forbidding child labor cannot be compromised. Companies collaborate with official fair-trade associations to guarantee that the workers are treated fairly and with respect (Joy et al., 2012; Pérez et al., 2022).

The rise of sustainable fashion is causing a fundamental transition in the fashion industry, which has long been criticized for its abusive labor conditions and environmental impacts. Sustainable fashion production methods are now a hub for innovation and constructive change since consumers worldwide are supposed to be more aware of the ethical issues, the rise of social and economic inequalities, and the environmental impacts of the fashion industry (Hur & Cassidy, 2019; Zollo, 2021).

2.2 *Sustainable Fashion Consumption*

The fashion industry is one of the oldest and largest industries that has undergone significant changes over time. Previously, fashion was a phenomenon where trends changed minimally and were standard rather than dynamic. However, consumers became interested in new fashion in the 1990s, which led to a rapid acceleration of fashion. With the emergence of the concept of fast fashion, the view that clothing items were something that should not be kept for a long time was shaped. In addition, the concepts of excessive consumption and disposal were normalized (Valtanen, 2021).

Fashion consumption is driven by a number of factors, including individual preferences, social trends, and the strategies used by brands in their marketing activities. Fashion is directly linked to society, culture, and identity formation (Valtanen, 2021). In addition, sociocultural factors are intertwined with fashion consumption. Clothes represent an individual's identity, carry symbolic meaning to conform to or challenge social norms. All of this makes the consumption process multifaceted (Gazzola et al., 2020). But the truth is that the fashion industry, which constantly introduces new trends and thus creates a culture of continuous buying and throwing away, forces customers to always follow the new fashion, which contributes significantly to increased consumption (Valtanen, 2021).

Moreover, the emergence of online shopping platforms has opened a new era in fashion consumption, making it more accessible than ever. The virtual marketplace not only offers a comprehensive range of options but also encourages consumers to participate in the consumption process with unprecedented ease by providing a seamless shopping experience. As a result, the merging of digital platforms with the fashion industry has further intensified the frequency and volume of consumption (Gazzola et al., 2020).

The term "sustainable consumption" refers to the use of resources that do not harm the environment irreversibly or disrupt natural systems to support the ability of present and future generations to meet their needs (Jackson, 2004). It covers the goods that help society and the environment during production and consumption to shape a sustainable future (Wei & Jung, 2017). Sustainable consumption in fashion is a conscientious and responsible approach to selecting and using apparel items. This practice places a strong emphasis on thoughtful choices by both producers and consumers within the fashion industry. At its core, sustainable consumption encourages individuals to be mindful of their purchasing habits, prioritizing quality over quantity and endorsing brands that uphold transparent and ethical practices (Jestratijevic & Rudd, 2018). As sustainable consumption advocates the use of environmentally friendly materials and production processes and the promotion of ethical labor practices across the entire supply chain; by adopting sustainable consumption in fashion, individuals play a vital role in reducing waste and preserving valuable resources, and supporting the development of a more sustainable and ethical fashion industry.

In 2003, Fuad Luke introduced the idea of "Slow Design" based on slowing down the metabolism of human and Earth. At this point, "slow" is starting to be perceived as the antithesis of the idea of "fast" and as a treatment for its disadvantages. On the other hand, slow design is an argument for challenging and altering a design paradigm based on fast consumption today (Türkmen, 2009). Slow designs offer an alternative point of view that respects human living conditions, biological and cultural diversity, and the finite resources of the world; they are a reaction against the products and services that are promoted by the culture of consumerism (Can & Ayvaz, 2017).

"Slow fashion" is a movement in the fashion industry advocating for sustainability, responsibility, and ethical practices. Embracing slow fashion enables consumers to play a crucial role in promoting positive change in sourcing, production, and consumption practices. Slow fashion brands prioritize organic, recycled, or durable materials and ensure fair wages and improved working conditions for production. The importance of slow fashion lies in its response to the negative environmental and social impacts associated with fast fashion. Fast fashion's rapid production and disposable nature contribute to a significant carbon footprint, excessive waste, and labor exploitation. In contrast, slow fashion encourages responsible practices, aiming to reduce waste and support fair wages, ultimately empowering consumers to make conscious choices and invest in quality, timeless garments with a longer lifespan (Brewer, 2019). The clothing items created in the most basic styles with the best quality materials are examples of how slow design is employed in the collections. By considering what we purchase, who makes the product, and how this affects the product's quality, slow fashion promotes a sense of social responsibility (Can & Ayvaz, 2017).

Making thoughtful purchasing decisions, taking into account how fashion choices affect society and the environment, and endorsing companies that value ethical business practices are all part of fully embracing slow fashion. By doing this, people may influence good changes in sourcing, production, and consumption while also helping to create a more sustainable fashion sector. In order to minimize the personal apparel consumption and encourage eco-friendly behavior, including pre- and post-purchase behavior, slow fashion consumption entails pragmatic strategies based on anti-consumerist values. One of the most important strategies is the extension of the life of apparel by not disposing of clothing prematurely, repairing, and taking good care of it. Second-hand shopping either online or via second-hand shops significantly reduces the demand for new items. Hosting clothes swaps or exchanges reduces new buying by establishing a shared way of recycling and reusing clothing. Adopting the culture of renting or borrowing clothes reduces the emphasis on long-term ownership by promoting short-term use through rentals or borrowing from acquaintances. Upcycling or creative reuse, on the other hand, converts old clothing into completely different products or accessories, providing an eco-friendly solution to disposal. These strategies collectively aim to shift the prevailing mindset away from the continuous acquisition and disposal of clothing, fostering more conscious consumption practices contributing to the circular economy. By adopting these approaches, individuals contribute to a paradigm shift in the fashion landscape, aligning with principles that prioritize longevity, reuse, and conscientious choices in clothing consumption (Vesterinen & Syrjälä, 2022).

Although it is commonly used synonymously, there are distinctions between the concepts of "sustainable fashion consumption" and "slow fashion consumption." While sustainable fashion consumption seeks the environmental and ethical compliance of products and production processes, the basic characteristics of slow fashion consumption are to prefer fewer but necessary products, to buy timeless, quality clothes, to turn to classic pieces away from seasonal trends, to slow down consumption (reduce unnecessary shopping), to generally support local production, small brands and handcraft, to extend the life of products with options such as reuse, repair and recycle. The slow fashion movement actually emerged as a reaction against fast fashion. However, since slow fashion is a movement against the rapid production and disposable nature of fast fashion, which contributes to a significant carbon footprint, excessive waste and labor exploitation, it would be appropriate to say that it is no longer limited to consumption behavior, and that brands are focusing on environmentally and socially responsible materials and processes in their design and production processes, as is the case with sustainable fashion.

Alternative ways of conscious consumption also include mindful consumption. The essence of mindful consumption behavior is to demonstrate mindful behavior and make mindful decisions by understanding the consequences of the effects of consumption. This concept was first defined by Sheth et al. (2011) as "temperance in acquisitive, repetitive, and aspirational consumption at the behavior level, ensuing from and reinforced by a mindset that reflects a sense of caring toward self, community, and nature." With this definition, two basic dimensions that establish the relationship between sustainability and mindful consumption are revealed; mindful behavior and mindful mindset. Mindful consumption behavior includes avoiding unnecessary products, buying less, making fewer repetitive purchases, and not overconsuming, balancing consumption in a way that is consistent with one's values and beliefs and improves one's own well-being, as well as caring for community and nature. The essence of mindful consumption behavior is to exhibit mindful behavior and make mindful decisions by understanding the consequences of the effects of consumption (Lim, 2017; Sermboonsang et al., 2020). Caring about the consequences of consumption consists of individual local and global care (Butcher, 2022). This concept is based on the idea that individuals with mindful consumption behavior will direct their behavior towards sustainability, as they are more likely to make choices that are less harmful to the environment (Amel et al., 2009). Because it is stated that the idea of mindful consumption has a transformation and directing effect. Bahl et al. (2016) argued that if consumers pay accepting attention to bodily sensations, thoughts and emotions, they can make choices that are aligned with their needs, values and insights rather than blindly or habitually reacting in consumption.

Slow fashion, sustainable fashion, and mindful consumption are alternatives to disposable and excessive consumption habits. Slow fashion aims for craftsmanship and durability to disrupt the disposable culture of fast fashion, while sustainable consumerism promotes environmentally friendly and ethical practices. Strategies such as product durability, buying second-hand clothes, clothing swaps, job creation and recycling are realistic ways to reduce the industry's pollution footprint. Embracing slow and sustainable fashion can help push the industry towards greater environmental and ethical responsibility and a more sustainable and ethical future.

2.3 Sustainable Fashion Marketing

Marketing activities are of key importance in influencing consumer behavior, attitudes and beliefs. Therefore, an appropriate marketing strategy can promote sustainable behavior and drive consumers towards sustainable purchasing behaviors. The increase in social and environmental sensitivity has led institutions, organizations, and companies to a more responsible process and first of all, social responsibility projects have been implemented, then sustainability, a broader holistic approach, has begun to be integrated into business models. In this context, with the spread of the concept of sustainability, new approaches have been added to the marketing approach. Among these new approaches, it is possible to list concepts, whose essence is to provide balance in the triangle of people, planet and profit, as sustainable marketing, green marketing, social marketing and holistic marketing.

Sheth and Parvatiyar (2013) initially defined sustainable marketing as a marketing approach that promotes sustainable development and ecosystem protection. Further, components of the marketing mix with the three facets of sustainability (economic, social, and environmental) were combined to define sustainable marketing in a more comprehensive way (Quoquab et al., 2021). Thus, sustainable marketing is now defined as the integration of environmental, social, and economic concerns and practices into the strategic and marketing activities, and covers sustainable production and consumption, reduction of consumption and wastage, and development of resource-efficient ways of consumption (Gordon et al., 2011; Calvo-Porral, 2019). Likewise, green marketing orientation was defined as the degree to which a company uses internal, strategic, and tactical procedures and activities to develop, sell, and provide goods and services with as little negative influence on the environment as possible (Papadas et al., 2017). However, it is emphasized that this definition should include other important environmental issues such as energy consumption, resource depletion, species extinction or ecosystem destruction; therefore, it should refer to a much broader adoption of sustainability behavior, including product-related decisions to reduce the environmental footprint, environmentally friendly packaging, recyclable or reusable content, examination of the product life cycle and a reverse supply chain approach to recover the highest possible value of the product (Cronin et al., 2011; Calvo-Porral, 2019). Social marketing focuses on changing attitudes and promoting sustainable marketing solutions for the benefit of society as a whole (Calvo-Porral, 2019). The broad perspective of holistic marketing, which is a value-oriented approach, overlaps with the holistic perspective of sustainability. Holistic marketing suggests that all stages of marketing should be carried out in a responsible and sustainable manner, and it contributes to the economic, social, environmental and administrative dimensions of sustainability (Akça Güleş, 2024).

In line with all these approaches, brands in the fashion industry convey to consumers all the activities they carry out for sustainability. The most commonly used tool by fashion brands to reach and interact with consumers today is undoubtedly social media and marketing activities on these platforms. Recently, social media marketing has surpassed traditional marketing activities. Especially after COVID-19, influencer marketing has become very important for fashion brands. Many fashion brands regularly update their social media profiles with high-quality images and videos showcasing their latest fashion collections. Their content is designed to capture the attention of the target audience and inspire them to explore the brand further. In addition to creating brand awareness and keeping consumers updated by sharing new products, the focus of brands' social media marketing strategies is to ensure customer satisfaction and loyalty by constantly interacting with their audience. Engagement with the audiences covers the issues such as asking for reviews, encouraging the consumers to share their experiences with the brands, and providing prompt customer support (Vladimirova et al., 2024). Zara uses social media posts to create a visual catalog so that users can quickly access and interact with the newest styles in fashion. It uses influencers who are compatible with the brand and target audience values (Ruan et al., 2022).

Focuses on engaging with the audience in its social media strategies, Nike, one of the leading global sportswear brands, invites its fans to share their personal stories and experiences by utilizing various strategies, including hashtags like #nikefc, #nikewomen, and #justdoit, to start discussions and foster a sense of community. Besides, the brand informs its social media followers on a regular basis about its products in an effort to inform and engage them (https://keyhole.co/blog/nike-social-media-strategy/). Social media allows creative minds to promote their brands via various campaigns. Using personalization with the #HereToCreate hashtag campaign, Adidas was able to reach the followers of approximately 30,000 athletes by turning them into brand advocates and create a huge target audience (https://www.invisionapp.com/inside-design/adidas-films/).

This power and influence of social media also stand out as a means of promoting the sustainability activities of brands as well as increasing consumer awareness. Inviting individuals to bring unused clothes from any brand and in any condition to stores since 2013, H&M encouraged consumers to close the loop in fashion by recycling unwanted clothes at their stores with the #WorldRecycleWeek hashtag in April 2016 (https://sustainablebrands.com/read/press-release/why-social-media-and-sustainability-should-go-hand-in-hand; https://www.instagram.com/reel/BEVRSjDgURd/).

Marks & Spencer is another fashion brand that specializes in using social media to promote sustainability initiatives. The brand frequently shares content about environmentally friendly approaches, fair trade practices and initiatives to reduce waste on social media platforms to gain credibility and create a positive brand image among socially conscious consumers by highlighting its efforts in this field. The company launched Plan A 2025, a new sustainability plan, in 2017. Focusing on climate change, waste, resources, fair partnerships, and health issues, Plan A integrates customers and suppliers to become a net zero business across all its value chain, and many posts are shared with #PlanA and #SustainabilityStories hashtags (https://corporate.marksandspencer.com/sustainability).

To increase the awareness about social and environmental impacts of fashion, social media campaigns are organized as a part of fashion activism movements. The campaign organized by Fashion Revolution with the #WhoMadeMyClothes hashtag invited individuals to question labor conditions and ethical concerns for increasing the visibility of fashion workers and drawing attention to the concept of transparency in the supply chain. Many global fashion brands shared their colleagues who play a role in the production processes on their social media accounts with the slogan "I Made Your Clothes." To draw attention to reducing the environmental damage caused by the fashion industry by reducing clothing consumption, Fashion Revolution launched another campaign with the hashtag #LovedClothesLast. The campaign aimed to encourage consumers to share various methods to increase product life during use and disposal of clothing and to spread these habits in the community (Vladimirova et al., 2024). It is seen that social media serves as a platform for influencers not only to introduce products and trends but also to raise awareness about a more sustainable lifestyle. The concerns about overconsumption and unethical fashion practices have been addressed by social media influencers from the Zero Waste, Slow Fashion, or Minimalist movements (Ramjaun, 2021).

3 Social Marketing

3.1 Key Features and Importance of Social Marketing

The term social marketing was first defined in 1971 by Philip Kotler and Gerald Zaltman to address the societal concerns and enhancing public welfare (Brennan & Binney, 2010). To encourage socially beneficial behavior, commercial marketing concepts and strategies are used in social marketing. Therefore, social marketing aims desired change in the society through behavior change. Social marketing is motivated by the desire to further social good, as opposed to commercial marketing, which tries to sell goods or services. It encourages people to adopt habits that lead to beneficial social consequences by utilizing audience engagement, communication tactics, and behavioral scientific insights (Andreasen, 2003).

By going beyond the main goal of marketing, which is increasing sales and making more profit or simply promoting a product or service through marketing strategies, the purpose of social marketing is to educate individuals and society as a whole and ensure to adopt healthy behaviors. Social marketing can be used by governments or nonprofit organizations to prevent issues such as domestic violence, reduce harmful habits such as smoking or drug use, emphasize the importance of protecting the environment, and provide information and ensure to take action for it or raise the awareness of the society on issues concerning public health, for example, the necessity of protection to prevent sexually transmitted diseases or the importance of vaccination and their impact on public health (Kotler, 1970; Craig Lefebvre & Flora, 1988; Varey 2010; Storey et al., 2015).

The key components of social marketing are defined as follows (Andreasen, 2002; French et al., 2010; Almestahiri et al., 2017):

- ***Consumer/audience research:*** It involves identifying and understanding the characteristics, interests, behaviors, and needs of the target audience. To obtain a thorough grasp of people's daily lives, a variety of research investigations combine data from many sources, including the public and private sectors.
- ***Behavioral change goal:*** The primary goal of social marketing is to motivate a specific audience towards adopting certain attitudes or behavior. It seeks to initiate actual behavioral change, either promoting sustainability of the environment, preventing alcohol misuse, or embracing healthier eating. It's a behavior change approach and not a communication approach. It seeks to accomplish as well as maintain long-term behavior modification compared to the traditional communications. Effective social marketing starts with the identification of behavioral goals that is used to design and evaluate interventions, and subsequently succeeding by measuring the results through intervention measures and advertising campaigns.
- ***Segmentation and targeting:*** Segmentation is a process by which a population is divided into distinct segments based on characteristics that influence their reaction to marketing interventions. Segmentation of target audiences in social marketing is required for obtaining maximum efficiency and effectiveness by using scarce resources. Audience segmentation seeks the target of opportunity, or those most likely to adopt the behavior, in contrast to traditional intervention tactics that concentrate on those with the greatest need. It is more suitable to divide the audience based on the attitudes or behaviors that are most relevant to the target audience rather than demographic traits.
- ***Marketing mix:*** In the context of social marketing, the behavioral offer presented to the target audience is referred to as the product. Price is the cost that the

targeted audience will pay to engage in a particular behavior. Both monetary (direct) and nonmonetary (indirect) costs may be included in the price. The location is where the intended audience will carry out the desired action, purchase any material things, and obtain any related services. Persuasive communication activities intended to motivate the target audience to take action are referred to as promotion. Promotion is used to inform the target audience of the location and to emphasize the qualities and benefits of the desired behavior.

- ***Exchange***: It focuses on realizing what people must give up in order to obtain the suggested advantages. This entails assessing all expenses (financial, social, and time), contrasting perceived and real costs and benefits, and customizing rewards, disincentives, and incentives according to the values of the target audience.
- ***Competition:*** Other activities that compete with the targeted activity are referred to as competition in social marketing. Through competition, social marketers can provide advantages that best set their desired behaviors apart from those of their rivals and create a long-lasting competitive edge that increases the customer appeal of their products. Both internal and external competition should be examined and addressed. Internal ones involve psychological factors, pleasure, desire, and risk taking, while external ones, which promote and reinforce alternative or counter behavior, entail broader influences and influencers competing for the audience's time, attention, and behavior.
- ***Theory:*** Behavioral theories provide useful frameworks for the methodical planning and assessment of social marketing interventions. Social marketing takes into account behavioral theory in four key areas including biological and physical factors, psychological factors, social factors, and environmental/ecological factors.

Both social marketing and traditional marketing rely on the concepts of communication and persuasion in order to influence the target group. They involve research and planning to generate effective messages and target the right audience. Kotler and Lee (2008) state that traditional marketing and social marketing both utilize the same set of principles and methods such as 4P (product, price, place, and promotion). They want to sell but traditional marketing desires to sell a product/service whereas social marketing attempts to sell "a behavior" so it is defined a little more difficult as adopting a beneficial behavior or letting go of an undesirable one is something one voluntarily does.

The ability of social marketing to drive meaningful and positive change within local communities and society overall is what renders it so precious. Social marketing applies marketing strategies to address large social problems, as opposed to traditional marketing, which is aimed at product promotion. Social marketing campaigns are vital in social justice, environmental sustainability, and public health because they can impact attitudes, behavior, and perceptions. Not only do these programs increase knowledge, but they also provide individuals with the capability to make decisions that benefit society as a whole on an individual and a community level. Social marketing is an evolving tool for involving the public to raise awareness, encourage community participation, and create genuine, long-term change. Its relevance is found in how it uses persuasion and communication to instill constructive actions and behaviors, and as such, transform the world into an empathetic, responsible, and fair one (Weinreich, 2012).

3.2 *Social Marketing Appeals*

Social marketing appeals are used to promoting desired behavior and preventing undesired behavior for the betterment of society. Two types of appeals are generally mentioned in this regard; emotional appeals and rational appeals. Emotional appeals are those appeals that generate good or bad feelings among the consumer for inducing behavioral change. Emotional appeals can be positive (e.g. enthusiasm) or negative (e.g. fear). Positive appeals draw on different techniques of communication to encourage habits favorable to such ends, while negative appeals discourage individuals from habits likely to slow progress and harm society. Emotional appeals attempt to establish a state of high psychological activation that compels the individual to take an action. Rational appeals, however, are interested in the characteristics of the product and use consumer benefits as appeals. Rational appeals can function through reflective consideration of arguments within the appeal, potentially leading to a modification of behavior and belief (Noble et al., 2014). In social marketing campaigns and communication, emotional appeals convey messages that appeal to the emotions and persuade the target audience through emotional stimuli. Rational appeals, also called informative appeals, are designed to persuade the target audience through rational thought processes. Rational appeal campaigns are designed to provide information and facts without emotion and to persuade the target audience through appeals to reason and logic. Rational or informational appeals aim to reach the mind of the target audience by conveying a set of logical information relevant to them and presenting facts directly to appeal to reason, thought, and awareness without any emotion in the message. Providing information about the problem and offering solutions is the main feature of the informative/rational tone in advertising (Casais & Pereira, 2021). These kinds of appeals encourage individuals to do things that lead to public health, social welfare, environmental conservation, or other social objectives (Noble et al., 2014).

Through highlighting the negative effects of certain choices or actions, negative social marketing appeals attempt to dissuade individuals from adopting such decisions or actions. Negative appeals use fear, guilt, or shame to discourage certain activities from being pursued by negative appeals, as compared to positive appeals which attempt to promote desired action. Negative appeals in public health communications, for example, can remind the audience of the serious health effects of smoking or the risk of not using a seat belt (Antonetti et al., 2015).

Negative emotional appeals like guilt, shame, or fear are used to discourage people from doing something bad or wrong. Fear appeals have the ability to cause anxiety or worry through emphasizing the possible danger or penalty of doing something.

Guilt arguments cause people to assess their actions by putting emphasis on the moral values and social consequences (Antonetti et al., 2015). The purpose of inciting fear is to raise enough physiological arousal in consumers to cause them to alter their behavior and take action against the threat in order to get rid of their fear (Ruiter et al., 2001). Similarly, customers are motivated by guilt to "do the right thing" and change their behavior in order to avoid feeling guilty again (Baek & Yoon, 2017). Guilt appeals evoke an emotional distress and it is lessened if the behavior to be elicited is taken in an attempt to make things right (Brennan & Binney, 2010). In a study to examine the effectiveness of guilt appeals in the context of a social marketing advertisement for the reduction of fashion consumption, findings showed that feelings of guilt have a mediating impact on the intention to reduce fashion consumption. This means that a campaign incorporating a guilt appeal will be more effective in encouraging the intention to decrease clothing consumption by evoking guilt, compared to a campaign that does not utilize a guilt appeal (Schutter, 2020). Shame appeals aim to make people ashamed or embarrassed of themselves so that they can be made to withdraw behaviors which will cause society to exclude them (Antonetti et al., 2015). It is outside the individual's control and accompanied by feelings of worthlessness and self-dislike. Shame is related to avoidance coping strategies; people who are ashamed try to conceal and hide. Whereas, guilt involves feelings of responsibility, regret, and remorse and is associated with action coping strategies. Guilt is under the control of the individual and associated with a specific situation (Pounders et al., 2018). While these strategies are certainly effective at bringing attention to the issue and discouraging wrong behavior, a balance of the ethical issues must be ensured so that the affective influence is one of empowering change and not causing undue harm. While such appeals can generate awareness and prompt behavior change, care needs to be taken as they also have the ability to induce negative emotions that will prompt defensive or fearful responses. Developing the right balance between conveying the seriousness of negative consequences and generating empowerment towards positive change is of paramount importance in the ethical utilization of negative social marketing appeals (Antonetti et al., 2015).

By using both emotional and rational approaches in a balanced way, effective social marketing campaigns can both raise awareness and encourage positive change in sustainable fashion consumption.

3.3 *Social Marketing Activities of Fashion Brands*

The fashion industry has been at a crossroads in recent years as it reckons with the environmental and social impacts of its long-criticized practices. Although the fashion industry has long been criticized for its ecological footprint, it is currently undergoing a shift toward sustainability. As people slowly become more aware of the environmental and ethical impacts of what they wear, fashion brands are starting to reevaluate their business practices and shift their production processes into a more sustainable approach. As this awareness grows, the fashion world is changing its business practices. This shift relates not only to the way fashion brands produce clothing but also to the way they promote that clothing to customers (Jung et al., 2020).

Fashion brands engage in sustainable marketing through advocating for sustainable materials, promoting fair and transparency supply chains, fair wages and labor practices, recycling programs, and emphasis on product quality, durability, and longevity. The companies interested in sustainability not only seek to engage consumer awareness but also wish to positively impact society and the environment. These brands are seeking to lend consumer education, transparency, and innovation, in order to create a new normal within the industry while fostering more responsible consumer attitudes toward fashion consumption, and addressing demands for more sustainable products, and hence, shaping a greener future. However, in order to raise consumer awareness and create widespread behavioral change in society, it is necessary to utilize social marketing beyond sustainable marketing.

According to Heim, who analyses the social marketing campaigns of the textile industry using the motivation-opportunity-ability behavior (MOAB) model and exchange theory, there exists a significant distinction between "sustainability marketing" and "social marketing." Sustainability marketing primarily focuses on presenting product features and production processes to consumers, such as claims of being made of organic or recycled materials under environmentally friendly and ethical conditions, and it is used as a means of building brand reputation and credibility. Meanwhile, an important issue to be considered in sustainable marketing activities is that they can sometimes be misleading and lead to discussions about greenwashing. Greenwashing is a deceptive marketing strategy used by companies to present themselves as environmentally conscious falsely. It includes employing vague terms like "natural" or "ecology-friendly," making inaccurate statements, and downplaying major environmental concerns while highlighting less important effects. The aim is to attract environmentally conscious consumers, despite the company's potential lack of genuine commitment to sustainability. On the other hand, social marketing is geared towards educating consumers and fostering behavioral change. Social marketing campaigns may encourage consumers to engage in a sustainable lifestyle by promoting the eco benefits of or ethically sourced textiles and emphasizing the importance of recycling/reusing and promoting awareness of not overusing textiles-related products, which can help consumers within the textile lifecycle towards more sustainable and responsible choices in terms of purchasing/using textiles. In essence, while sustainable marketing offers products to consumers, that is, it applies a push strategy, social marketing aims to make consumers demand sustainable products due to behavioral change, that is, it applies a pull strategy. Undeniably, social marketing campaigns may not always achieve the goal completely. Though social marketing is a new marketing strategy for fashion brands, it can lead to an increase in brand perception and ultimately an increase in consumption. On the other hand, although social marketing does not change consumer behavior, it triggers public discourse and opinion. In this way, social marketing can still be effective in changing fashion consumer behavior to improve social justice in the fashion industry (Heim, 2022).

When social marketing campaigns of fashion brands are examined, for example, Patagonia was admittedly the first to run a social marketing campaign for the fashion industry with its powerful running campaign "Don't Buy This Jacket." Initially, that campaign revealed the impact of overconsumption in the clothing industry practices. At Patagonia, "Don't Buy This Jacket" became a remarkably successful campaign—not only did it address overconsumption in the fashion industry but also the campaign graduated to a new sense of sustainable marketing for fabrics— where customers bridged the importance of sustainability with all other textiles (cotton, polyester, etc). It can be said that this campaign, which was initially launched as a social marketing campaign, transformed into a sustainable product campaign. Regardless of that, the brand not only aimed to promote low consumption through this campaign but also managed to bring together activist groups working towards environmental issues for a common purpose (Heim, 2022).

A Spanish fashion brand Ecoalf's business strategy is founded on the circular economy, a framework intended to maximize resource utilization and reduce waste. Using waste, Ecoalf creates high-end clothing instead of virgin resources. These wastes include fishing nets, plastic bottles, used tires, marine debris, and even coffee grounds. By turning waste materials into high-performance textiles, Ecoalf lessens the carbon footprint of the fashion sector while simultaneously contributing to environmental cleanup. One of Ecoalf's most prominent projects is "Upcycling the Oceans," an initiative launched in 2015. This project, run in collaboration with the Ecoalf Foundation, partnered with local fishermen to rid the ocean of plastic waste. Fishermen, mostly in Spain and later in Thailand, collect tons of plastic and other waste from the ocean, which Ecoalf turns into yarn for fabric. What started as a local initiative has evolved into an international collaboration with plans to expand as awareness of ocean pollution grows. The company states in its advertising visuals which materials it recycles and uses in textile and apparel products (Calvo-Porral, 2019; Zott, 2024).

By romanticizing quality and longevity, a Spanish designer brand Adolfo Dominguez launched a campaign called "Old Clothes" for Spring/Summer 2020. The campaign included photos of models wearing the brands old clothes donated by the consumers. All photos had a bold headline claiming how old the clothes were like "This Trench is 30 Years Old," "This Suit is 40 Years Old," and "This Coat is 30 Years Old". An accompanying video campaign featured interviews with donors of clothing, who said they prefer "one good outfit over three bad ones" and "a basic wardrobe that can be worn for years to come" (https://consciousfashion.co/guides/sustainable-fashion-marketing). The designer supports this campaign with "Repeat More, Need Less" slogan (https://www.instagram.com/adolfodominguezofficial/p/CFoicmDi-L5/). This campaign actually emphasized the unnecessary nature of excessive consumption and that one can always be stylish with basic pieces.

Another social marketing campaign was launched by famous denim brand Levi's in 2021. The "Buy Better, Wear Longer" campaign suggested to its customers to use the product they bought for many years, to buy second-hand clothing and to extend the life of the clothes by using tailors; in the words of brand president Jen Sey, Levis denim was a product designed to be worn for generations. Heim states that influencers and rising idols participated in the multifaceted international campaign, including Marcus Rashford, a young football player for Manchester United and the England National Team. The advertisement, developed by the creative company AKQA, sought to promote Levi's as a fast-fashion alternative while bringing attention to the negative environmental effects of the production and consumption of clothing and 75% of consumers perceive sustainability to be crucial yet the creative agency disagreed with the opinion that consumers bought the products they liked, thus they eventually bought them again. After all, the consumer is encouraged to think of a more expensive, more durable object with emotional attachment (https://www.levistrauss.com/2021/04/22/levis-launches-buy-better-wear-longer-campaign/).

Nike's "Move to Zero" campaign has introduced the Space Hippie shoe models, which is a tremendous step forward in sustainable social marketing in fashion. The campaign is in place to promote shopping that is responsible and ethical, highlighting the implications of excessive shopping on our limited earth resources while promoting sustainable alternatives. Launched as a social campaign, "Move to Zero" sparks awareness and showcases Nike's tangible commitment to reducing consumers' ecological footprint through sustainable product offerings. Importantly, each Space Hippie silhouette incorporates 25–50% recycled content and Nike is oriented towards sustainable practice. The campaign serves the dual purpose of introducing shoes but also represents Nike's larger intent in creating positive change in the fashion industry by working with activist groups focused on consequential public environmental issues (https://www.nike.com/sustainability).

Tentree has initiated effective cause-related marketing, both to support their products and raise awareness around more environmentally sound choices. Tentree is sharing knowledge via social media as an educational tool to demonstrate how material science and sustainable sourcing of fabrics can support consumers at large. The Seafleece campaign in early 2023 is a notable example. The Seafleece demonstrates Tentree's commitment to saving and protecting the oceans, because it constructs fabrics from recycled plastics and uses resources to combat the environmental impacts of ocean plastic waste. Tentree is supporting environmentally aware consumers, but is also pushing a broader environmental agenda, drawing attention to the power of social media campaigns to create impact in the textile industry. One of the products reflective of sustainability is Seafleece. Seafleece total composition is 60% recycled cotton and 40% REPREVE® Our Ocean® Recycled Polyester—this medium-stretch knit fabric delivers luxurious, soft-to-the-touch textile that is also supportive of sustainability. Seafleece is made with recycled waste, including recyclable bottles, textile waste and fishing nets, inviting consumers to enjoy comfort and warm styles while creating a greener planet and supporting Tentree's sustainable practice (https://www.tentree.com/pages/sustainability).

In 2020, Renzo Rosso, president of the OTB group and founder of Diesel, engaged in a partnership with Eco-Age to create initiatives for sustainability and inclusivity. This resulted in the project for "Responsible Living," which introduced the "Library" project. The "Library" strategy contained a campaign that showed the various types of denim working together to form the recognizable Diesel logo. The

S/S 2022 Library collection would concentrate on denim because it is Diesel"s core business. The "Library" collection will contain garments made from gender-neutral and sustainable denim. This new denim collection would provide responsibly sourced materials and focus on reducing water and chemicals. In a push for transparency, consumers can access sourcing information through a QR Code. Half of the Library collection is designed to be seasonless and offers a wardrobe aesthetic, supporting Diesel's focus on sustainability and inclusivity in fashion (https://global.diesel.com/for-responsible-living/).

The iconic foldable bag, Le Pliage from Longchamp that many people know so well due to its colors and sizes, reinvented itself with a green focus in 2021. This was part of a campaign for the brand promoting changing waste into something useful with the plastic from a trash can to represent a message of waste recycling. The bags in this collection are made from recycled nylon with a CO_2 emissions impact of production over six times less than that of a pair of denim. The ultimate goal is to use recycled materials throughout the collection by 2022. The successful sustainability makeover of Le Pliage takes on a new level of desirability while preserving the practicality and aesthetics of the bag. (https://www.longchamp.com/us/en/maison-longchamp/sustainable/recycled-materials.html).

4 Consumer Identity and Sustainable Consumption

4.1 Self-Identity and Sustainable Consumption

Self-identity means an individual's self-conception, self-referential cognitions, or self-definition that people apply to themselves as a result of the roles they occupy or their regular particular behavior (Biddle et al., 1987). Self-identity is explained as a part of an individual's personality that is linked to a certain behavior and makes him or her unique (Conner & Armitage, 1998). It acts as a label that a person uses to adopt or indicate a certain behavior.

Self-identification demonstrates how a person's perspective on role-appropriate behavior can influence his or her place in society (Sharma et al., 2020). The concept of self-identity is included within the scope of marketing research because it comprises attitudes and judgments that may affect consumers' decisions to purchase products. Regarding consumption behavior, people consume products not only to satisfy their needs but also to reflect themselves in front of others (Bagozzi et al., 2013). In other words, self-identity represents how fashion consumers view themselves and influence their selection of brands that align with their desired social image.

Moreover, green self-identity is defined as an individual's self-perception of being environmentally conscious (Van der Werff et al., 2013). Many studies have been conducted evaluating the role of self-identity on sustainable purchase intention (Legere & Kang, 2020; Patel et al., 2020; Judge et al., 2019; Mishra et al., 2023;

Cao & Yang, 2022). It was noted that individuals with a strong sustainable self-identity exhibit environmental protection behavior (Bryan et al., 2013). People are motivated to maintain a clean, green environment not due to social norms or laws, but because of a deeper desire to preserve and strengthen their self-identity (Lee et al., 2016).

The relationships between materialism, environmental concern, social consumption motivation, and sustainable consumption behaviors are largely or totally mediated by pro-environmental self-identity (Dermody et al., 2015). Besides, the expression of self-identity was found as a key driver of intention to use reusable products in retail stores (Lin et al., 2022). Ultimately, when individuals see themselves as committed and consistent green consumers, their intention of green consumption is more likely to convert into actual purchases of green products. Their self-identity reinforces and aligns with their intentions, increasing the likelihood that they will act in ways that reflect their green values and beliefs (Hong et al., 2024). Nevertheless, environmentally and socially conscious self-identity, characterized by a deep concern for environmental and social issues, was found to have a negative impact on eco-fashion purchase intention, which was attributed to the consumers' adoption of buying less and higher prices commanded by eco-fashion products (Lambert, 2019).

In summary, self-identity refers to how individuals perceive themselves, influencing their consumption choices. People often buy products not just for functionality but also to express their identity. A strong green self-identity motivates individuals to engage in environmentally friendly behaviors, such as purchasing sustainable fashion. Research shows that self-identity can be a key factor in turning green intentions into actual purchases. However, some consumers with a strong environmental identity may avoid eco-fashion due to high prices and a preference for minimal consumption.

4.2 *Social-Identity and Sustainable Consumption*

Social identity theory, as a social psychology theory, states that individuals define their sense of self-using social categories or group memberships and provides explanations for intra-group dynamics, inter-group relations, and the collective self. It was put forward by Tajfel (1979) based on the idea that social identity is shaped through social interactions and communicative behaviors. While self-identity refers to certain behaviors and characteristics of an individual, social identity is derived from the behaviors that occur due to the social groups to which individuals belong (Brown, 2000).

Social identity theory suggests that people develop or desire social identities to go beyond their personal identities to express and form their own selves (He et al., 2012). The identity of an individual is created by these social interactions and evaluations (Guan & So 2016). Social identity consists of three components: cognitive, affective, and evaluative. Cognitive social identity refers to the perception that individuals are part of a specific social group. Affective identity involves the emotions linked to being a member of that group. Evaluative identity relates to the positive or negative judgments people form about the group they belong to, also known as collective self-esteem (Jiang et al., 2022).

Consumer behavior is often driven by the need to enhance both self-identity and social identity (Tian et al., 2001). Given the significant role these identities play in consumption choices, researchers have explored their influence on sustainable consumption behavior (Bartels & Reinders, 2016). Since social interaction serves to affirm an individual's self-identification, the development of one's self-identity and social identity occur simultaneously (Stets & Burke, 2000). This can also be identified as such that sustainable fashion consumers express two interconnected identities in consumption. The individual level, where it serves as a means of self-identity expression, and the socio-relational level, where it reflects their social identity. While establishing their self-identities, individuals also shape their social identities to align with a group and avoid rejection or to distinguish themselves from others. Ethical consumers try to strike the best possible balance between their self-identity and social image. Because sustainable fashion items will satisfy their demands for self-definition and social image, customers who have a strong ethical self-identity—that is, those who believe they are ethical consumers—will be more inclined to sustainable fashion consumption (Niinimäki, 2010).

On the other hand, studies suggest that consumers with a strong social-identity are more likely to adopt sustainable consumption behaviors than those with a stronger self-identity, as such behaviors allow them to enhance their social image by demonstrating their commitment to environmental concerns (Griskevicius et al., 2010). Additionally, sustainable consumption behavior gives consumers a sense of belonging to society, enabling them to easily integrate into desirable social groups. This sense of identification arises because consumers believe their consumption choices can provide specific social benefits, which further strengthens their social self-concept (Abdulrazak & Quoquab, 2018). Moreover, social influence exerted by others can shape consumer decisions, encouraging them to choose environmentally friendly products over conventional alternatives (Salazar et al., 2013). Conventional fashion consumption often stems from an individualistic and self-interested identity orientation, like purchasing a coat for its functionality. However, sustainable fashion consumption is driven by relational motives by a concern of others, and collectivistic motives for promoting societal well-being (Mukendi et al., 2020).

In another study investigating the sustainable consumption behavior of different genders based on salient identity, the findings indicated that gender influenced sustainable consumption when self-identity was more prominent, with female participants exhibiting higher levels of sustainable behavior than their male counterparts. However, when social identity was emphasized, both male and female participants displayed similar levels of sustainable consumption. In this context, male participants increased their sustainable behaviors to match those of female participants, likely driven by social comparison and the desire for recognition. This implies that when social identity was salient, engaging in sustainable behavior may have served as a symbolic marker of distinction (Costa Pinto et al., 2014).

Social identity is shaped by group membership and social interactions. It influences consumer behavior by encouraging people to adopt sustainable practices to enhance their social image. Moreover, social influence plays a crucial role in motivating sustainable choices. On the other hand, although studies show that consumers with a strong social identity are more likely to engage in sustainable consumption than those with a stronger self-identity, as it helps them integrate into desirable social groups, ethical consumers often balance their self-identity with social expectations. As a result, consumer identity plays a significant role in sustainable consumption, with self-identity driving personal choices and social identity shaping behavior through societal expectations.

In connection with self-identity and social identity, sustainable and ethical consumption can be triggered in marketing with appeals targeting self-benefit and other-benefit. Marketing strategies that balance self-benefit and other-benefit appeals can effectively influence sustainable consumer behavior. These concepts will be discussed with examples of sustainable fashion in Sect. 4.3.

4.3 Self-Benefit and Other-Benefit Marketing Appeals

In marketing and advertising activities aimed at encouraging consumers to make charitable donations and ethical or sustainable consumption, the messages focus on the benefits of the behavior for others in society or the individual himself or herself. The messages that focus on the benefits of the consumer itself refer to self-benefit appeal, whereas the messages that focus on the benefits of others in society refer to other-benefit appeals. These appeals are also defined as egoistic (self-benefit) and altruistic (other-benefit) appeals (Ryoo et al., 2020).

While some studies have found that altruistic appeals provide more motivation toward ethical or sustainable consumption, some studies have favored egoistic appeals. In ethical or sustainable consumption, it can be considered that other-benefit appeals can be effective when there are environmental considerations, whereas self-benefit appeals can be effective when the individual's health or financial interests and material savings are at stake (Van den Berg, 2020). Additionally, the effectiveness of marketing appeals is influenced by an individual's personal interest, materialism level, self-awareness and whether they prioritize collective or personal identity (Yang et al., 2015; Ryoo et al., 2020; Rask, 2021).

Empirical evidence on the attitudes of young consumers towards self-benefit and other-benefit marketing appeals captures their respective impacts on sustainable consumption decision-making. Rask (2021) clarifies that such appeals are not required to introduce new knowledge but only serve to remind consumers of the ecological costs of consumption. The self-benefit appeal consolidates some aspects of sustainable consumption, while the other-benefit appeal provides general understanding of the complexities in sustainable consumption. Hence, the other-benefit appeal is more effective in generating overall awareness. However, survey findings indicate that self-benefit appeals have a greater impact on young consumers' preference and choice of sustainable products as they like products offering personal benefits rather than collective benefits.

Similarly, a study on fair trade jeans (Ryoo et al., 2020) found that highly materialistic consumers responded more positively to ethical product advertisements when their self-interests were emphasized. This suggests that individuals with a strong focus on personal gain are more likely to engage in sustainable consumption when the messaging aligns with their priorities.

The effectiveness of abstract (unclear) versus concrete (tangible) appeals also depends on how green product benefits are perceived and is influenced by public self-awareness and identity orientation (Yang et al., 2015). Abstract appeals are more persuasive than concrete ones when the product benefits others, particularly for individuals with high public self-awareness or a collective identity. However, when the product benefits oneself, both appeal types appear to be less effective, and the distinction between abstract and concrete messaging becomes insignificant.

In the dynamic realm of marketing and advertising, staying attuned to consumer motivations and preferences is imperative for successful brand positioning and engagement. These two prominent approaches that have gathered significant attention in recent years, which are self-benefit and other-benefit marketing appeals, play pivotal roles in shaping consumer perceptions, influencing decision-making processes, and fostering lasting connections between brands and their target audiences (Shan et al., 2023).

Self-benefit appeals are known for their focus on the individual in a world of infinite wants. Companies are able to convince the consumer that his or her life is definitely getting better with the product or service being purchased by using self-benefit appeals. Marketers, based on Maslow's hierarchy of needs, meet the individuals' self-actualization need by portraying their products as an integral part of happiness and a tool for personal growth (Stephen, 2023). This was supported by the findings of the study introducing that a benefit appeal targeting certain market segments, for instance, the health and wellness or luxury product sectors, has the potential of giving the highest performance boost while customization emerges at the top as key drivers.

As compared to self-benefit appeals, other-benefit appeals are driven by the sense of serving a broader good. Such an approach highlights the ways in which a certain product or service positively adds to the well-being of others or to society in general. Capitalizing on feelings of social obligation and moral motivation, other-benefit appeals resonate with individuals' impulse to improve society. Brands that are skilled at strategically leveraging other-benefit appeals position themselves alongside causes for charity or advocate green sustainability. This positioning generates a strong sense of partnership with the consumer, creating a more intense relationship based on a shared interest in good social effect (White & Peloza, 2009).

Consumer decision-making is indeed a multifaceted process influenced by cognitive, emotional, and social factors. The utilization of self-benefit appeals in advertising tends to resonate with individualistic motives, tapping into consumers' desires for personal fulfillment and gratification. Conversely, other-benefit appeals have been found to evoke emotional responses, promoting a sense of social connectedness and responsibility among consumers. This understanding allows for the creation of more effective and compelling advertising strategies that align with the diverse psychological and cultural inclinations of consumers (Shan et al., 2023).

Modern marketers rely on self-benefit and other-benefit marketing appeals as crucial tools. It is important for marketers to grasp the psychological and ethical aspects of these strategies to create effective campaigns appealing to various consumer groups. Striking a balance between self-gratification and social responsibility enables marketers to establish meaningful connections beyond simple transactions, fostering lasting brand loyalty. While self-gratification or self-satisfaction is related to self-identity about creating or strengthening one's characteristics, social responsibility or the desire to create a social image is triggered in connection with social identity. The relationships of all these concepts affecting sustainable consumption are shown in Fig. 1.

Considering the use of self-benefit and other-benefit appeals in the fashion industry, for example, Zara took advantage of both appeals when introducing its sustainable collection, Join Life. The care for fiber highlighted the self-benefit due to the use of organic fibers and consumer health, and the care for water and the planet highlighted the other-benefit due to the use of environmental-friendly production technologies (Wei & Jung, 2022).

H&M highlighted the other-benefit appeal in the collection described as the next level of sustainable denim that was produced together with American denim brand, Lee by using 100% recycled cotton, water-saving dyeing processes and more sustainable choices for all details (for example, 100% recycled polyester sewing thread) (https://www.hm.com/om/lee-x-hm/).

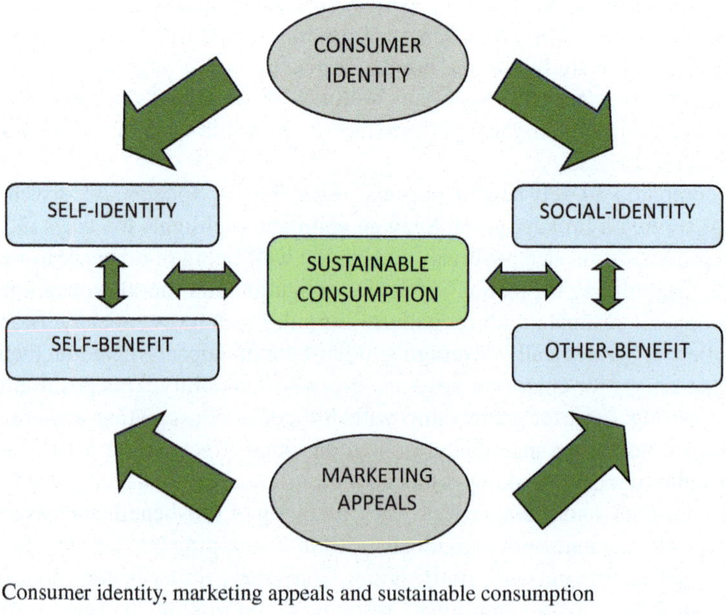

Fig. 1 Consumer identity, marketing appeals and sustainable consumption

Marketing messages promoting ethical or sustainable consumption often focus on either self-benefit (egoistic appeal) or other-benefit (altruistic appeal). Self-benefit appeals highlight personal gains such as health or financial savings, while other-benefit appeals emphasize social and environmental impact. While some studies favor altruistic appeals, others suggest egoistic appeals can be more effective in specific industries. Successful brands strategically use both approaches in their sustainability campaigns to connect with diverse consumer motivations.

5 Empirical Study

5.1 Research Model and Data Analysis

The literature survey on the effects of social marketing on consumer research revealed that a group of studies examined the role of advertising contents and labels containing negative emotional appeals for products that threaten public health in changing consumer behavior. The contents of the studies carried out for this purpose is given below.

Boshoff (2021) investigated the effectiveness of fear and guilt-fear appeals on behavioral change of consumers through warning labels on high-sugar-content products. The purpose was to investigate emotional and cognitive responses of consumers to the negative emotional appeals. However, since each consumer's readiness for behavioral change may be at different levels, they were categorized into three groups as individuals at, pre-contemplation, contemplation, and preparation stages. In the field study carried out with the participation of 460 people in South Africa, the results revealed that the cognitive responses of each consumer group were similar, that is the use of negative emotional appeals did not result in enhanced positive cognitive responses compared to no appeal. However, different negative emotional appeal warning labels affected consumers' emotional responses differently. It was concluded that social marketers can use fear appeals for consumers in the pre-contemplating and those in the preparation stage, and guilt-fear appeals for consumers in the contemplation stage.

For exploring the long-lasting positive effect of shock advertising on quit smoking, Mukattash et al. (2023) conducted a survey with a focus group of 41 smoking participants in Jordan. Participants were exposed to children images including a blend of emotional and fear appeals, and it was found that designing shocking advertisings that create guilt, fear, and sadness can lead to designing shocking adverts that behavioral changes for avoiding harming other people, in particular children due to second-hand smoking.

Yousef et al. (2021) examined the influence of positive (love, joy and friendship), negative (fear, guilt, shame and sadness) appeals and their combinations on young adults' drink driving perceptions. Theory of Reasoned Action (TRA) was used as a predictive model, and the results revealed that while positive emotional appeals lead to significantly lower drink driving attitudes than negative appeals, negative emotional appeals significantly changed the participants' social norms. When positive and negative emotional appeals were combined, participants had significantly fewer positive attitudes, lower social norms, and reduced intentions to drink and drive.

Although the use and effects of fear appeal have been extensively examined in studies on social marketing, specifically the effect of guilt and shame appeals as negative self-conscious emotions on behavior change has been also observed. Temel (2023) investigated the effects of public spots containing self-conscious messages such as guilt and shame on behavioral change. The researcher hypothesized that the message framing method (behavior-focused or self-focused) can use guilt and shame in different ways to elicit behavior change intention. The basic assumption of the model investigating the behavioral change was that guilt-inducing stimuli in the messages lead to the intention to change behavior as a result of eliminating the suppression, whereas shame-inducing stimuli in the messages lead to the intention to change behavior as a result of activation of suppression. The study also emphasized that the use of fear appeal in changing public attitudes is limited to issues directly related to oneself, regardless of the harm that may be caused to others, but self-conscious emotions direct the individual's attention from focusing on oneself to interpersonal and group relationships.

The roles of guilt and shame appeals were also investigated in health communications depending on individuals' self-construal in terms of eating healthy, responsible drinking, quitting smoking, and staying-at-home intentions during the COVID-19 pandemic. It was concluded that guilt appeals were more persuasive and related to the individuals with an independent self-construal, whereas shame appeals were more convincing and related to the individuals with an interdependent self-construal (Chan & Septianto, 2024).

As a result of the positive effects of social marketing, especially on health-related behavior change, the effectiveness of the tool in question was also examined on sustainable and ethical consumption. Due to increasing environmental issues, the effect of social marketing on sustainable consumer behavior has been analyzed through various models. In charity advertising, it was found that individuals who experienced more guilt were more likely to participate in ethical consumptions and making donations (Xu, 2022). Lunden et al. (2020) used expanded Theory of Planned Behavior (TPB) with positive and negative emotional appeals referring to social marketing to examine its effects and accuracy in predicting intention to avoid palm oil consumption. Regarding the effect of negative emotional appeals, the findings revealed that most of the participants felt varying levels of guilt as response to the negative emotional appeal that was significantly higher than for the positive emotional appeal. Another important finding in the research was that the core constructs of TPB (attitude, subjective norm and perceived behavioral control) had a positive and direct impact on intention, and the effect of these core constructs were stronger when the participants were exposed to negative emotional appeals.

The study conducted by Van Breda et al. (2023) explored the impact of fear and guilt appeals on sustainable consumption among South African millennials. Using a quantitative approach with a between-subject experimental design based on TPB, the research founded out that fear and guilt appeals in sustainability advertising did not significantly alter millennials' pro-environmental intentions. On the other hand, attitude had the strongest effect on behavioral intention. The study is focused on investigating pro-environmental behavior in general, rather than on a specific product or industry.

Balderjahn and Hoffmann (2024) analyzed the impact of different social marketing appeals on sustainable consumption, focusing on informative, social normative and emotional appeals. The study assessed their efficacy in reducing spending, examining negative and positive rebounds, and promoting sustainable consumption patterns. Within the scope of the study, a survey was conducted with 170 participants in a grocery in Germany and the analysis proved that social normative and emotional appeals were effective in reducing shopping expenditures, revealing their potential to influence consumer behavior towards sustainability. Additionally, informational and social normative appeals increased donations, whereas emotional appeals reduced gifts received. Specifically, Schutter's (2020) research on fashion consumption found that a guilt-inducing advertising increases feelings of guilt, which in turn strengthens the intention to reduce clothing consumption compared to an advertising that does not evoke feelings of guilt.

In addition to the models based on the TPB model and expanded by using emotional appeals as an antecedent in the research of sustainable consumption behavior, Cao and Yang (2022) expanded the TPB model by adding social identity and self-identity constructs in their study and examined the mediating effect of these additional constructs on behavioral intention of commuting consumers towards mobile as a service (MaaS). According to the results, sustainable identity was the most important part for consumers, indicating that consumers attached great importance to self-value affirmation and mutual approval of sustainable concepts.

In another study exploring the effect of COVID-19, social identity, and socially responsible food consumption on generations, Leyva-Hernández et al. (2023) stated that social identity positively and significantly influenced socially responsible consumption. Perceived harshness, perceived advantages, and cue to action all positively and significantly influenced social identity. Furthermore, there is a characteristic called identity that acts as a whole mediator between perceived advantages and socially conscious consumption, perceived seriousness and socially conscious consumption, and action cue and socially conscious consumption. Perceived barriers only directly affect consumption that is socially responsible. In the same way, there are differences in the ways that taking action, social identity, and membership to a social network group relate to each other for Generations X, Y, and Z.

When the studies were examined, it was seen that negative emotional appeals are persuasive in changing individuals' behavior, consumers' self and social identities are effective in sustainable consumption behaviors. In this context, the following model (Fig. 2) was built and relevant hypotheses were established to examine the impact of social marketing on sustainable consumption regarding fashion consumption, which is one of the consumptions that creates the most environmental burden in the world. The conceptual model was based on the Stimulus—Organism—Response (SOR) model, which provides a basic theoretical framework for

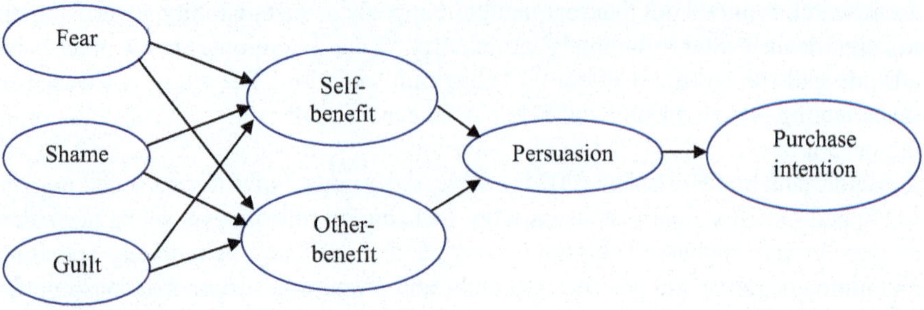

Fig. 2 Conceptual framework

understanding consumer behavior. Social marketing appeals referred to stimuli, self-benefit and other-benefit appeals referred to organism, and finally persuasion corresponded to response in the SOR model. The constructs of the model were defined below:

Fear: Fear appeals emphasize the dangers of engaging with and ignoring a specific product, service, or concept, warning of bad consequences if one does not comply. They rely on creating a sense of threat to a person's well-being, which prompts them to take action to manage or prevent negative consequences (Williams, 2012).

Guilt: Although guilt appeals are commonly used in persuasive communication to encourage selfless deeds, their results are not always guaranteed. While guilt usually motivates people to take constructive actions, it can also have negative effects such as reactance (Graton & Mailliez, 2019).

Shame*:* Shame appeal is used for persuasion that attempts to change behavior by causing individuals to feel ashamed or embarrassed about their actions. Its goal is to trigger feelings of inadequacy or guilt, prompting people to modify their behavior to avoid or address these negative emotions (Alibwaini & Ünal, 2022).

Self-Benefit: Self-benefit appeals involve persuasive messages emphasizing the individual advantages or rewards one could gain from taking a specific action, such as donating (White & Peloza, 2009).

Other-Benefit: Other-benefit marketing appeals emphasize the additional advantages of a product or behavior beyond its main function, utilizing emotional connections, social validation, or self-expression to encourage adoption. They highlight factors like emotional resonance, social validation, or self-expression to persuade individuals to embrace the promoted behavior or product (White & Peloza, 2009).

Persuasion: Persuasion is the process of convincing people to adopt specific views, attitudes, or actions by employing strategies such as logical reasoning, emotional appeal, credibility, or social standards, with the goal of influencing their thoughts or behavior in the desired direction (Isabella, 2022).

Purchase Intention for Sustainable Fashion: It describes an individual's inclination to purchase environmentally conscious clothing items, indicating their planned or anticipated shopping actions in favor of sustainable fashion. This intention is shaped by personal values, attitudes, and the perceived advantages associated with sustainable clothing.

Based on the conceptual model, the following hypothesis were established:

H1: Fear appeal has a significant influence on self-benefit of the individual.
H2: Shame appeal has a significant influence on self-benefit of the individual.
H3: Guilt appeal has a significant influence on self-benefit of the individual.
H4: Fear appeal has a significant influence on other-benefit of the individual.
H5: Shame appeal has a significant influence on other-benefit of the individual.
H6: Guilt appeal has a significant influence on other-benefit of the individual.
H7: Self-benefit appeal has a significant influence on persuasion of the individual.
H8: Other-benefit appeal has a significant influence on persuasion of the individual.
H9: Persuasion has a significant influence on purchase intention of the individual.

After the conceptual model was built, 3 visuals were designed to encourage sustainable fashion consumption before preparing the survey (Fig. 3). It was aimed to trigger feelings of fear, guilt and shame in consumers through these visuals. For this reason, a pilot study was conducted to check whether the designed visuals evoke the relevant emotion. The pilot study was conducted with 32 participants. The ages of the participants ranged between 18 and 45. Participants with different profiles, such as university students and employees, were reached through an online survey. Within the scope of the pilot study, 3 out of 9 emotions (scared, anxious, uncomfortable, tense, nervous, blameworthy, remorseful, guilty, embarrassed, ashamed, humiliation) were asked to be selected for each visual. The most frequently chosen emotions were anxious, blameworthy and scared for Fig. 3a, blameworthy, guilty, anxious for Fig. 3b, and blameworthy, guilty, and ashamed for Fig. 3c, respectively. In this case, since scared, guilty and ashamed feelings were chosen for different visuals, it was decided that the visuals were suitable and the study continued with the main field work.

After the pilot study was carried out, the suitability of the images was ensured and the survey preparation for the main study was completed. The questionnaire for the main survey was built to involve two parts. In the first part, demographic characteristics of the participants such as age, gender, education, and income level were asked. The second part included 28 questions (indicators) to evaluate the role of social marketing on promoting sustainable fashion consumption. 11 indicators were adapted to evaluate the feelings of the participants that the visual evokes about fashion consumption, 3 indicators were adapted to evaluate the self-benefit, 3 indicators were adapted to evaluate the other-benefit, 5 indicators were adapted to evaluate the persuasion, and 6 indicators were adapted to evaluate the purchase intention for sustainable fashion. Survey items were structured on a 5-point Likert scales (1 = strongly disagree, 5 = strongly agree).

The data set was analyzed using Smart-PLS software. Based on the conceptual framework, the constructs of the measurement model were identified and after evaluating the reliability and validity of the measurement model by using confirmatory

Fig. 3 Visuals were designed to encourage sustainable fashion consumption

composite analysis (CCA), a structural model was created for hypothesis testing to test the structural relationships with the help of partial least squares structural equation modeling (PLS-SEM).

5.2 *Results and Discussion*

The data was collected from 300 participants, whose demographic characteristics were given in Table 1. As seen in Table 1, 75% of the participants are female and 24% are male. When we look at the age distribution, it can be seen that the majority are 24 years old and younger, members of Generation Z. In the questionnaire, four income categories that were specified as low, low to medium, medium and medium to high income classes were included. 60% of the participants are undergraduate students and hence most of the participants have low level of personal income that is 10.000 TL or under.

For evaluating the data, analyses of the measurement model of the study were carried out by using CCA to assess whether it was suitable for the PLS-SEM used to perform hypothesis tests.

As a first step of a reflective measurement model assessment in CCA, loadings of indicators for each construct were evaluated. Constructs and indicators with their references and loadings were given in Table 2. Loadings should have a value of at least 0.708 and an associated t-statistic above ±1.96 to be significant for a two-tailed test at the 0.05 significance level (Hair et al., 2011). According to the results, one indicator (Q21) with a factor loading of 0.584 was removed and the rest 27

Table 1 Sociodemographic characteristics of the participants

		Frequency	%
Gender	Female	225	75.0
	Male	72	24.0
	Prefer not to say	3	1.0
Age	24 and below	263	87.7
	25–44	31	10.3
	45–59	5	1.7
	60 and above	1	0.3
Level of education	High school attending or graduate	69	23.0
	Attending undergraduate	180	60.0
	Bachelor graduate	42	14.0
	Enrolled in graduate school	7	2.3
	Graduate from graduate school	2	0.7
Personal income level	Low	213	71.0
	Low to medium	44	14.7
	Medium to high	23	7.7
	High	20	6.6

Table 2 Constructs and indicators

Constructs and indicators	References	Loadings
Fear		
Q3—When I see these picture ads, I feel anxious.	Becheur and Valette-Florence (2014)	0.888
Q4—When I see these picture ads, I feel uncomfortable.	Becheur and Valette-Florence (2014)	0.858
Q1—When I see these picture ads, I feel tense.	Becheur and Valette-Florence (2014)	0.854
Q5—When I see these picture ads, I feel scared.	Becheur and Valette-Florence (2014)	0.809
Q2—When I see these picture ads, I feel nervous.	Becheur and Valette-Florence (2014)	0.760
Guilt		
Q10—When I see these picture ads, I feel blameworthy.	Becheur and Valette-Florence (2014)	0.887
Q11—When I see these picture ads, I feel remorseful.	Becheur and Valette-Florence (2014)	0.873
Q9—When I see these picture ads, I feel guilt.	Becheur and Valette-Florence (2014)	0.866
Shame		
Q6—When I see these picture ads, I feel embarrassed.	Becheur and Valette-Florence (2014)	0.818
Q7—When I see these picture ads, I feel ashamed.	Becheur and Valette-Florence (2014)	0.875
Q8—When I see these picture ads, I feel humiliation.	Becheur and Valette-Florence (2014)	0.791
Self-benefit		
Q12—These picture ads reminded me that we need to think about ourselves.	Kim et al. (2021)	0.878
Q14—These picture ads reminded us that we should focus on caring and improving our own lives.	Kim et al. (2021)	0.898
Q16—These picture ads evoked the need to look after one's own interests.	Kim et al. (2021)	0.719
Other-benefit	Kim et al. (2021)	
Q13—These picture ads reminded me that we need to think about others.	Kim et al. (2021)	0.878
Q15—These picture ads reminded us that we should focus on caring and improving others' lives.	Kim et al. (2021)	0.898
Q17—These picture ads evoked the need to look after others' interests.	Kim et al. (2021)	0.809
Persuasion		
Q18—I think the visual is persuasive.	Becheur and Valette-Florence (2014)	0.895
Q19—I think the visual is convincing.	Becheur and Valette-Florence (2014)	0.850

(continued)

Table 2 (continued)

Constructs and indicators	References	Loadings
Q20—From now on, I will be more receptive to this type of information about fashion consumption.	Becheur and Valette-Florence (2014)	0.902
Q22—The visual made me stop and think.	Becheur and Valette-Florence (2014)	0.877
Purchase intention for sustainable fashion		
Q23—I will consider switching to eco-friendly brands for ecological reasons.	Maichum et al. (2016)	0.882
Q24—I intend to purchase green products next time because of its positive environmental contribution	Maichum et al. (2016)	0.910
Q25—I plan to purchase more green products rather than normal products.	Maichum et al. (2016)	0.870
Q26—I plan to make my clothes last longer by recycling them or giving them away/selling them.	Zhang (2014)	0.841
Q27—I would consider purchasing recycled fashion items.	Chaturvedi et al. (2020)	0.866
Q28—I would consider purchasing second-hand fashion items.	Adeyemi (2023)	0.707

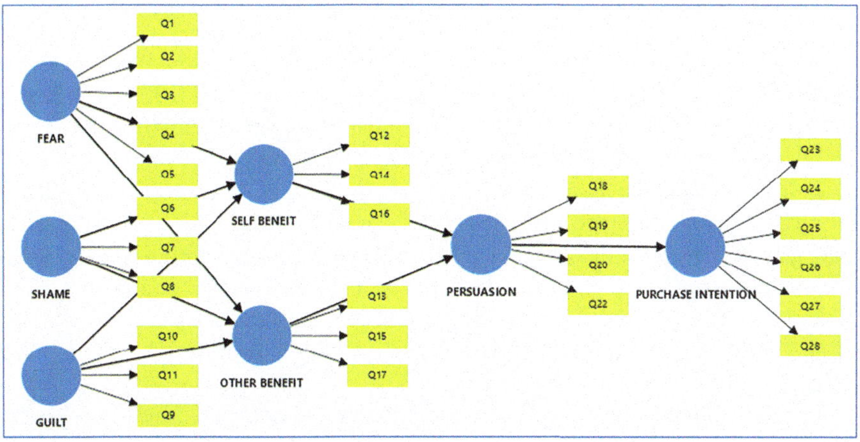

Fig. 4 Measurement model

indicators with the loadings ranged between 0.707 and 0.910 yielded seven factors. The indicator Q28 had a loading of 0.707, which is on the cutoff value. However, since some researchers consider values of 0.60 and above as appropriate (Dash & Paul, 2021), the indicator in question was not removed from the model. The measurement model with the constructs and the indicators were shown in Fig. 4.

Within the scope of analyses regarding the evaluation of the measurement model; internal consistency analyses including Cronbach Alpha (CA) and Composite Reliability (CR); convergent validity analyses including Average Variance Explained (AVE) and discriminant validity analyses including Cross-loading, Fornell-Larcker, and Heterotrait-Monotrait Ratio (HTMT) were conducted.

Table 3 Construct reliability measures

	Cronbach Alpha (CA)	Composite Reliability (CR)
Fear	0.891	0.920
Guilt	0.849	0.908
Shame	0.777	0.868
Self-benefit	0.787	0.873
Other benefit	0.830	0.897
Persuasion	0.904	0.933
Purchase intention	0.921	0.939

Table 4 Convergent validity measures

	Average Variance Extracted (AVE)
Fear	0.697
Guilt	0.766
Shame	0.687
Self-benefit	0.698
Other benefit	0.744
Persuasion	0.777
Purchase intention	0.720

As seen in Table 3, CA values of all constructs of the measurement model are between 0.777 and 0.921, and CR values are between 0.868 and 0.939. As a rule of thumb, reliability measures should be above 0.70 (Hair et al., 2020), and the reliability of the constructs were proved with the obtained values.

In the next step, the assessment of convergent validity was performed by calculating the Average Variance Extracted (AVE) for all constructs. AVE is calculated as the mean of the squared loadings of each indicator associated with a construct, and shows the average variance shared between the construct and the related indicators. If the AVE value is above 0.50, it indicates that convergent validity is achieved (Hair et al., 2020). As seen in Table 4, convergent validity was achieved since all AVE values are above 0.5.

For the evaluation of discriminant validity, cross-loadings, Fornell-Larcker criterion, and HTMT were assessed. Discriminant validity is measured to check whether the constructs in the model are unrelated (Voorhees et al., 2016; Eroğlu, 2023).

The cross loadings analysis results were given in Table 5. In cross loadings analysis, a particular indicator should have higher loadings on its own parent construct in comparison to other constructs in the study. If an indicator loads well onto another construct in comparison to its own parent construct, then there are issues of discriminant validity (Hair et al., 2014). In the present study, all indicators were accurately aligned with their respective factors, as shown in the table.

As another indicator of discriminant validity, Fornell-Larcker analysis results were seen in Table 6. Here, the values in the diagonal must be the highest of all values in the column they are in (Hair et al., 2021). The values in the diagonal are the square root of AVE values, and the other values are the correlation coefficients between constructs. When the values in Table 6 are examined, the AVE square root

Table 5 Cross loadings

	Fear	Guilt	Shame	Self-benefit	Other-benefit	Persuasion	Purchase Intention
Q1	0.854	0.494	0.479	0.311	0.515	0.520	0.478
Q2	0.760	0.473	0.475	0.185	0.376	0.468	0.363
Q3	0.888	0.577	0.528	0.278	0.467	0.540	0.403
Q4	0.858	0.573	0.573	0.266	0.437	0.594	0.443
Q5	0.809	0.504	0.516	0.270	0.447	0.431	0.386
Q6	0.554	0.706	0.818	0.116	0.278	0.482	0.354
Q7	0.572	0.712	0.875	0.118	0.357	0.469	0.365
Q8	0.424	0.442	0.791	0.308	0.341	0.331	0.239
Q9	0.522	0.866	0.694	0.191	0.368	0.463	0.374
Q10	0.601	0.887	0.585	0.252	0.464	0.579	0.459
Q11	0.516	0.873	0.648	0.282	0.374	0.430	0.374
Q12	0.271	0.226	0.180	0.878	0.430	0.301	0.358
Q13	0.507	0.456	0.406	0.448	0.878	0.433	0.411
Q14	0.313	0.307	0.240	0.898	0.542	0.347	0.402
Q15	0.524	0.403	0.346	0.536	0.898	0.458	0.524
Q16	0.191	0.123	0.154	0.719	0.403	0.184	0.225
Q17	0.340	0.322	0.259	0.451	0.809	0.313	0.363
Q18	0.550	0.472	0.394	0.321	0.430	0.895	0.527
Q19	0.430	0.422	0.415	0.273	0.355	0.850	0.461
Q20	0.614	0.521	0.459	0.303	0.434	0.902	0.614
Q22	0.543	0.564	0.503	0.317	0.442	0.877	0.596
Q23	0.460	0.443	0.339	0.357	0.433	0.563	0.882
Q24	0.420	0.432	0.343	0.340	0.462	0.568	0.910
Q25	0.415	0.388	0.298	0.365	0.403	0.534	0.870
Q26	0.480	0.417	0.341	0.301	0.443	0.563	0.841
Q27	0.448	0.383	0.330	0.392	0.478	0.539	0.866
Q28	0.302	0.275	0.257	0.328	0.352	0.417	0.707

Table 6 Fornell-Larcker analysis results

	Fear	Other-benefit	Persuasion	Purchase Intention	Guilt	Self-benefit	Shame
Fear	0.835						
Other-benefit	0.541	0.863					
Persuasion	0.612	0.474	0.881				
Purchase intention	0.500	0.507	0.629	0.849			
Guilt	0.627	0.463	0.565	0.463	0.875		
Self-benefit	0.318	0.554	0.346	0.408	0.279	0.835	
Shame	0.614	0.399	0.505	0.377	0.728	0.235	0.829

value of each construct (0.835 for fear, 0.863 for other-benefit, 0.881 for persuasion, 0.849 for purchase intention, 0.875 for guilt, 0.835 for self-benefit, 0.829 for shame) is higher than the correlation coefficients in its column. Therefore, the Fornell-Larcker discriminant validity conditions were met.

Table 7 HTMT analysis result

	Fear	Other-benefit	Persuasion	Purchase Intention	Guilt	Self-benefit	Shame
Fear							
Other-benefit	0.610						
Persuasion	0.675	0.533					
Purchase intention	0.545	0.572	0.681				
Guilt	0.718	0.538	0.634	0.516			
Self-benefit	0.364	0.680	0.391	0.464	0.317		
Shame	0.749	0.475	0.610	0.452	0.924	0.269	

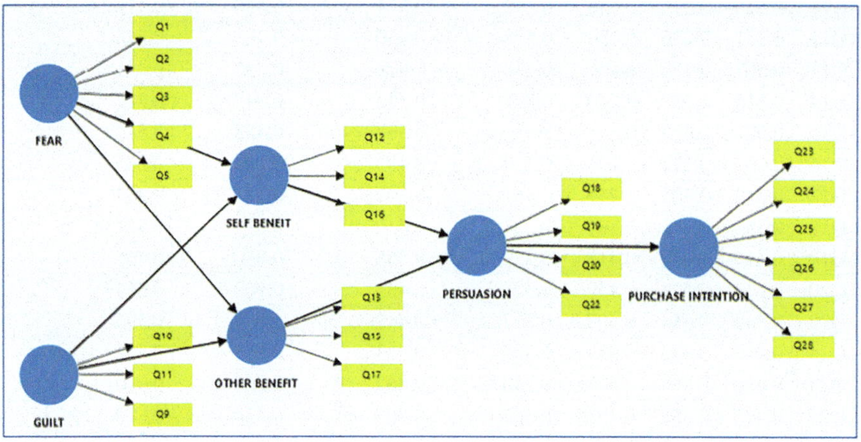

Fig. 5 Final measurement model

As the third criterion of discriminant validity, HTMT was checked. The HTMT analysis results are shown in Table 7. HTTP ratio expresses the ratio of the average of the correlations of the indicators of all constructs in the research to the geometric means of the correlations of the indicators of the same construct (Henseler et al., 2015). The HTMT coefficient should be below 0.90 if the constructs being measured are theoretically close to each other and below 0.85 for more distinct constructs (Hair et al., 2020). When the obtained HTMT values are examined, it is seen that the correlation between the "shame" and "guilt" constructs is high. Therefore, the "shame" construct was removed from the measurement model and the analysis continued with the final measurement model seen in Fig. 5. The structural model for PLS-SEM analysis was created based on the final measurement model.

As a first step of structural model assessment, multicollinearity was checked. If multicollinearity exists in the model, it means the correlation between the constructs is very high. The criterion used to determine whether there is a multicollinearity problem is VIF (Variance Inflation Factor) values. For formative measurement models, external model VIF values (Outer Model VIF Values) are examined, and for

Table 8 Inner VIF values

	Fear	Other-benefit	Persuasion	Purchase Intention	Guilt	Self-benefit
Fear		1.806				1.806
Other-benefit			1.444			
Persuasion				1.000		
Purchase intention						
Guilt		2.394				2.394
Self-benefit			1.444			

Table 9 Hypothesis testing results

Hypothesis		Path coefficients	p values	t values	Result
H1	Fear → Self-benefit	0.237	0.001	3.394	Accepted
H3	Guilt → Self-benefit	0.130	0.049	1.971	Accepted
H4	Fear → Other-benefit	0.414	0.000	6.343	Accepted
H6	Guilt → Other-benefit	0.203	0.003	2.992	Accepted
H7	Self-benefit → Persuasion	0.120	0.047	1.983	Accepted
H8	Other-benefit → Persuasion	0.408	0.000	6.043	Accepted
H9	Persuasion → Purchase intention	0.629	0.000	11.020	Accepted

reflective measurement models, internal model VIF values (Inner Model VIF Values) are examined. Since the measurement model in this research are reflective, internal model VIF values were examined. It is stated that if the internal model VIF values are less than 5, the multicollinearity does not exist (Hair et al., 2020). Since the VIF values in Table 8 are less than 5, there is no multicollinearity problem in the model.

The hypothesis testing in the present study conducted through path analysis to observe the structural relations between the constructs. The path coefficients and the p values were given in Table 9. The path coefficients obtained in path analysis vary between -1 and $+1$. Values close to $+1$ indicate a strong positive relationship between variables, and values close to -1 indicate a strong negative relationship. At a 95% confidence interval, p values below 0.05 show that the relationship is significant. According to the preset p-value threshold of 0.05, all hypotheses were accepted.

When the path coefficients were examined, it was seen that all relationships were positive and the strongest relationship was between persuasion and purchase intention (Fig. 6). When the relationship of self-benefit and other-benefit constructs with persuasion is compared, a stronger relationship of other-benefit emerges. When social marketing appeals are taken into consideration, it is noted that the strongest relationship is between fear and other benefit. In other words, triggering the feeling of fear in the designed visuals mostly affected the motivation to do something for others.

Another criterion used to evaluate whether path coefficients are significant is the *t* value. When a *t* value is greater than the critical value, it is concluded that the path coefficient is statistically significant within a certain margin of error. The critical *t* value used in the 95% confidence interval for the two-tailed test is 1.96 (Hair et al.,

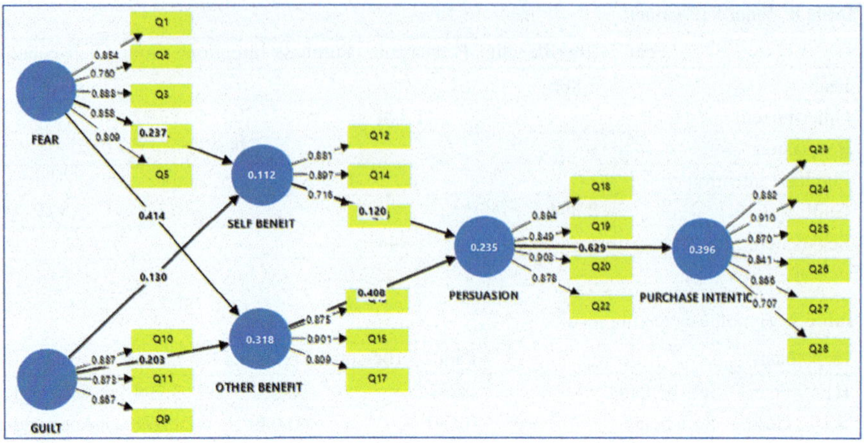

Fig. 6 Measurement and structural model

2017). As seen in Table 8, all seven paths had *t* values greater than the threshold value of 1.96.

The study offers valuable perspectives on the statistical significance and intensity of correlations among the assessed variables, considering both p-values and coefficient values. Acceptance of seven hypotheses suggests robust evidence of significant relationships. These findings contribute to a deeper understanding of the intricate dynamics involved in constructing a model to analyze the how social marketing appeals, such as fear, shame, and guilt, influence consumers' motivations and attitudes towards sustainable fashion.

The effect of each construct on purchase intention of sustainable fashion can be evaluated by R^2 value of each construct, which shows the accuracy of the model's prediction (Hair et al., 2017). In Fig. 6, it is seen that fear and guilt explain 11.2% of the self-benefit, and 31.8% of the other-benefit. Self-benefit and other-benefit explain 23.5% of persuasion, and persuasion explains 39.6% of purchase intention of sustainable fashion.

There are debates in the literature about whether it is meaningful to use goodness of fit indices, which are used to determine the fit of the model in SEM, in PLS-SEM models. Substantially, it was stated that evaluating goodness of fit indices in PLS-SEM is only useful when the purpose of the research is based on a purely confirmatory approach (Hair et al., 2020). When that is the goal, in the first step, overall model fit should be assessed using estimated model fit indices. Then, as a part of CCA, the measurement model is tested based on the saturated model fit indices. While the saturated model assesses the correlation between all constructs of the model, the estimated model takes into account the model structure (Henseler et al., 2016). However, PLS-SEM is based on causal-predictive relations since it maximizes the explained variance of the constructs in the model. Thus, purpose of confirmatory analysis can ignore the predictive features of the model (Hair et al., 2020). Consequently, for assessing the fit of the structural model, three criteria including

Standardized Root Mean Square Residual (SRMR), Squared Euclidean Distance (d_ULS), and Geodesic Distance (d_G) were calculated and evaluated for the saturated model.

SRMR is calculated by dividing the fitted residual values by the standard error of the residuals. It can take values between 0 and 1, and the lower the value is, the better the model fit (Dash & Paul, 2021). The threshold values of up to 0.08 is recommended (Hu & Bentler, 1999). The SRMR value was obtained as 0.061 that falls within an acceptable range. Moreover, Squared Euclidean Distance (d_ULS) and Geodesic Distance (d_G) provided evidences for the external validity of the constructs at 95% confidence interval, with the values of 1.106 and 0.575, respectively (Dijkstra & Henseler, 2015).

The present study presented out that fear and guilt appeals have positive impacts on individuals' self-benefit and other-benefit towards sustainable clothing purchasing behavior. The increasing awareness on the environmental issues of fashion consumption, even accelerated after COVID 19 pandemic positively affected consumers' habits and intentions to purchase more sustainable products for both themselves and others.

Compared to guilt appeal, fear appeal was found to be more effective on self-benefit. This was expected because fear appeals tend to drive self-beneficial behaviors when individuals perceive a personal threat and believe in their capacity to address it. Conversely, guilt appeals often encourage other-beneficial behaviors by invoking feelings of social responsibility and the desire to alleviate guilt through actions that help others (Seva, 2022). On the other hand, both fear appeal and guilt appeal were found to have more impact on other-benefit than self-benefit. Studies examining the impact of fear appeals, which use the emotion of fear to encourage people to adopt or avoid a certain behavior, have generally focused on health communication and public health or safety in traffic. At this point, it is normal for self-benefit motivation to come to the fore. However; when it comes to the damages in the environment and nature, it is inevitable for the consumer to think about other people and even future generations beyond himself or herself. Guilt appeal, which aims motivating individuals by making them feel guilty for not exhibiting or exhibiting a certain behavior, is effective in areas such as charity and social responsibility campaigns and environment and nature protection campaigns, for example, to get people to change their behaviors that harm the environment. Therefore, it is ordinary for guilt appeal to trigger other-benefit motivation more in the field of sustainable fashion consumption. As Baek and Yoon (2017) noted, people who feel guilty about not adopting a green lifestyle aim to repair the perceived damage they have caused by changing their behavior. Guilt motivates consumers to take remedial actions for others (Kim et al., 2021).

As a result of the empirical study, it was realized that both self-benefit and other-benefit motivations persuade the consumers for sustainable fashion consumption. This finding is consistent with the finding of a previous study concluded that organic cotton apparel consumption is generally driven by both egoistic and altruistic considerations (Han & Park-Poaps, 2024); therefore, as the authors pointed out, pro-environmental behavior of the consumers cannot be limited to their consideration of other-benefit. However, the result of the empirical study revealed that other-benefit was more effective on persuasion than self-benefit. The lack of dominance of self-benefit motivation may be related to the fact that the images used do not indicate features such as health benefits or product quality for sustainable fashion products. The greater effectiveness of other-benefit motivation may be attributed to an individual's awareness and sensitivity to how others perceive them. Individuals who try to behave in accordance with the expectations of others in social environments and tend to control their image, can aim to enhance their social image by portraying themselves as thoughtful consumers through appeals that emphasize benefits for others (Min et al., 2025).

5.3 *Practical Implications and Recommendations*

The environmental burden and social impacts created by the fashion industry have inevitably forced manufacturers and brands to transform. From raw materials to distribution, companies' efforts to reduce environmental impact and improve ethical practices are increasing day by day, and sustainable alternatives are gaining importance. On the other hand, these transformation steps taken towards sustainability can progress faster and be more effective if the preferences of end consumers change. Especially during and after the COVID-19 pandemic, as consumer environmental awareness and consciousness has increased, brands' promotion of their sustainability initiatives has become one of the most important components of their marketing activities. In this context, it can be said that social marketing, which educates consumers and supports behavioral change, is an important tool rather than traditional marketing. Many brands are now trying to encourage consumers for sustainable fashion consumption through social marketing campaigns. However, to maximize the effects of the campaigns, companies need to understand which social marketing appeals engage consumers' self- or other-benefit motivations and are therefore most likely to lead to behavior change.

The findings of the empirical study revealed that social marketing campaigns first stimulate fear and then guilt in consumers' mind. The outcomes that emphasized the higher impact of these appeals on other-benefit of consumers rather than self-benefit can help social marketers in building strategies for targeting consumers, who see sustainable consumption as an opportunity to strengthen their social image by showing others the importance they place on the environment. According to the results, both self-benefit and other benefit have positive effects in convincing the consumer towards sustainable consumption, but consumers are more convinced of sustainable consumption as a result of considering the benefits of others. Companies may therefore be advised to prioritize activating altruism and biospheric values in their social marketing campaigns, as these values enable consumers to focus more on their choices and the consequences of those choices on other people and the environment. Otherwise, sustainable consumption may not be sufficiently encouraged when there is a greater focus on hedonic and egoistic values.

The present study includes detailed information about the production, consumption and marketing of sustainable fashion, and examines the importance of social marketing, which, unlike traditional marketing, is more effective in changing consumer habits, and its applications in the fashion industry. The method and research model used in the study provides a framework to determine the effects of sustainable fashion campaigns on different target audiences (generation, gender, geography … etc).

Acknowledgement The authors would like to thank to Ayşe Nur DUNDAR, Elif AYTAS, Halil Sevban DANISMAN, Elif Sema AVSAR, and Sena Nur EMER for their contribution in conducting the survey of empirical study as a part of their Senior Design Project.

References

Aakko, M., & Koskennurmi-Sivonen, R. (2013). Designing sustainable fashion: Possibilities and challenges. *Research Journal of Textile and Apparel, 17*(1), 13–22.

Abdulrazak, S., & Quoquab, F. (2018). Exploring consumers' motivations for sustainable consumption: A self-deterministic approach. *Journal of International Consumer Marketing, 30*(1), 14–28.

Adeyemi, B. A. (2023). *The impact of demographic variables and sustainability awareness on purchase intention on second-hand clothing in Nigeria*, MSc Thesis, University of Gävle.

Akça Güleş, S. (2024). *The effect of sustainability on brand image: a research on y and z generations*, Doctoral Thesis, Ege University, İzmir, Turkey.

Alibwaini, I., & Ünal, G. (2022). An integration of self-conscious emotions into cognitive-functional model: The case of guilt and shame. *Current Research and Reviews in Psychology and Psychiatry, 2*(2), 169–183.

Almestahiri, R. D., Rundle-Thiele, S., Parkinson, J., & Arli, D. (2017). The use of the major components of social marketing: A systematic review of tobacco cessation programs. *Social Marketing Quarterly, 23*(3), 232–248.

Amel, E. L., Manning, C. M., & Scott, B. A. (2009). Mindfulness and sustainable behavior: Pondering attention and awareness as means for increasing green behavior. *Ecopsychology, 1*(1), 14–25.

Andreasen, A. R. (2002). Marketing social marketing in the social change marketplace. *Journal of Public Policy & Marketing, 21*, 3–13.

Andreasen, A. R. (2003). The life trajectory of social marketing: Some implications. *Marketing Theory, 3*(3), 293–303.

Antonetti, P., Baines, P., & Walker, L. (2015). From elicitation to consumption: Assessing the longitudinal effectiveness of negative emotional appeals in social marketing. *Journal of Marketing Management, 31*(9–10), 940–969.

Baek, T. H., & Yoon, S. (2017). Guilt and shame: Environmental message framing effects. *Journal of Advertising, 46*(3), 440–453.

Bagozzi, R. P., Verbeke, W. J., Dietvorst, R. C., Belschak, F. D., van den Berg, W. E., & Rietdijk, W. J. (2013). Theory of mind and empathic explanations of Machiavellianism: A neuroscience perspective. *Journal of Management, 39*(7), 1760–1798.

Bahl, S., Milne, G. R., Ross, S. M., Mick, D. G., Grier, S. A., Chugani, S. K., et al. (2016). Mindfulness: Its transformative potential for consumer, societal, and environmental wellbeing. *Journal of Public Policy & Marketing, 35*(2), 198–210.

Balderjahn, I., & Hoffmann, S. (2024). The effectiveness of consume-less appeals in social marketing. *Journal of Macromarketing, 44*(2), 258–275.

Bartels, J., & Reinders, M. J. (2016). Consuming apart, together: The role of multiple identities in sustainable behavior. *International Journal of Consumer Studies, 40*(4), 444–452.

Becheur, I., & Valette-Florence, P. (2014). The use of negative emotions in health communication messages: Study of the effects of fear, guilt, and shame. *Recherche et Applications en Marketing (English Edition), 29*(4), 89–109.

Behera, B. K. (2007). Comfort and handle behavior of linen-blended fabrics. *AUTEX Research Journal, 7*(1), 33–47.

Biddle, B. J., Bank, B. J., & Slavings, R. L. (1987). Norms, preferences, identities and retention decisions. *Social Psychology Quarterly*, 322–337.

Boshoff, M. (2021). *The influence of fear-and guilt-fear appeals on consumers in different stages of change: A study on high-sugar-content products*. Doctoral dissertation, Stellenbosch: Stellenbosch University.

Brennan, L., & Binney, W. (2010). Fear, guilt, and shame appeals in social marketing. *Journal of Business Research, 63*(2), 140–146.

Brewer, M. K. (2019). Slow fashion in a fast fashion world: Promoting sustainability and responsibility. *Laws, 8*(4), 24.

Brown, R. (2000). Social identity theory: Past achievements, current problems and future challenges. *European Journal of Social Psychology, 30*(6), 745–778.

Bryan, C. J., Adams, G. S., & Monin, B. (2013). When cheating would make you a cheater: Implicating the self prevents unethical behavior. *Journal of Experimental Psychology: General, 142*(4), 1001.

Butcher, C. (2022). *Understanding mindful consumption: A consumer perspective*. Doctoral dissertation, Open Access Te Herenga Waka-Victoria University of Wellington.

Calvo-Porral, C. (2019). The role of marketing in reducing climate change: An approach to the sustainable marketing orientation. In *Climate change and global development: Market, global players and empirical evidence* (pp. 261–283). Springer.

Can, Ö., & Ayvaz, K. (2017). Tekstil ve modada sürdürülebilirlik. *Akademia Doğa ve İnsan Bilimleri Dergisi, 3*(1), 110–119.

Cao, M., & Yang, C. (2022). Research on consumer identity in using sustainable mobility as a service system in a commuting scenario. *Systems, 10*(6), 223.

Casais, B., & Pereira, A. C. (2021). The prevalence of emotional and rational tone in social advertising appeals. *RAUSP Management Journal, 56*, 282–294.

Chan, E. Y., & Septianto, F. (2024). Self-construals and health communications: The persuasive roles of guilt and shame. *Journal of Business Research, 170*(114), 357.

Charter, M., Pan, B., & Black, S. (Eds.). (2023). *Accelerating sustainability in fashion, clothing and textiles*. Routledge.

Chaturvedi, P., Kulshreshtha, K., & Tripathi, V. (2020). Investigating the determinants of behavioral intentions of generation Z for recycled clothing: An evidence from a developing economy. *Young Consumers, 21*(4), 403–417.

Chi, T. (2015). Consumer perceived value of environmentally friendly apparel: An empirical study of Chinese consumers. *The Journal of The Textile Institute, 106*(10), 1038–1050.

Conner, M., & Armitage, C. J. (1998). Extending the theory of planned behavior: A review and avenues for further research. *Journal of Applied Social Psychology, 28*(15), 1429–1464.

Costa Pinto, D., Herter, M. M., Rossi, P., & Borges, A. (2014). Going green for self or for others? Gender and identity salience effects on sustainable consumption. *International Journal of Consumer Studies, 38*(5), 540–549.

Cowan, K., & Kinley, T. (2014). Green spirit: Consumer empathies for green apparel. *International Journal of Consumer Studies, 38*(5), 493–499.

Craig Lefebvre, R., & Flora, J. A. (1988). Social marketing and public health intervention. *Health Education Quarterly, 15*(3), 299–315.

Cronin, J. J., Smith, J. S., Gleim, M. R., Ramirez, E., & Martínez, J. D. (2011). Green marketing strategies: An examination of stakeholders and the opportunities they present. *Journal of the Academy of Marketing Science, 39*(1), 158–174.

Dash, G., & Paul, J. (2021). CB-SEM vs PLS-SEM methods for research in social sciences and technology forecasting. *Technological Forecasting and Social Change, 173*(121), 092.

Dermody, J., Hanmer-Lloyd, S., Koenig-Lewis, N., & Zhao, A. L. (2015). Advancing sustainable consumption in the UK and China: The mediating effect of pro-environmental self-identity. *Journal of Marketing Management, 31*(13–14), 1472–1502.

Dijkstra, T. K., & Henseler, J. (2015). Consistent and asymptotically normal PLS estimators for linear structural equations. *Computational Statistics & Data Analysis, 81*, 10–23.

Enes, E., & Saygili, B. (2023). Zero waste and modular fashion design: A suggested collection. *Textile and Engineer, 30*(131), 180–191.

Eroğlu, D. T. (2023). *Kişisel Satış Yaklaşımlarının Satış Performansına Etkisi Uzerine Bir Saha Araştırması*. Doctoral dissertation, Bursa Uludag University.

French, J., Blair-Stevens, C., McVey, D., & Merritt, R. (2010). Key concepts and principles of social marketing. Social marketing and public health: Theory and practice, 29–44.

Foksowicz-Flaczyk, J., & Walentowska, J. (2008). Eco-friendly antimicrobial finishing of natural fibres. *Molecular Crystals and Liquid Crystals, 484*(1), 207–573.

Gazzola, P., Pavione, E., Pezzetti, R., & Grechi, D. (2020). Trends in the fashion industry. The perception of sustainability and circular economy: A gender/generation quantitative approach. *Sustainability, 12*(7), 2809.

Gordon, R., Carrigan, M., & Hastings, G. (2011). A framework for sustainable marketing. *Marketing Theory, 11*(2), 143–163.

Graton, A., & Mailliez, M. (2019). A theory of guilt appeals: A review showing the importance of investigating cognitive processes as mediators between emotion and behavior. *Behavioral Sciences, 9*(12), 117.

Griskevicius, V., Tybur, J. M., & Van den Bergh, B. (2010). Going greento be seen: Status, reputation, and conspicuous conservation. *Journal of Personality and Social Psychology, 98*, 392–404.

Guan, M., & So, J. (2016). Influence of social identity on self-efficacy beliefs through perceived social support: A social identity theory perspective. *Communication Studies, 67*(5), 588–604.

Gwilt, A., & Pal, R. (2017). Conditional garment design for longevity. In *PLATE Conference, Delft, November 8–10, 2017* (Vol. 9, pp. 149–153). IOS Press.

Hair, J. F., Howard, M. C., & Nitzl, C. (2020). Assessing measurement model quality in PLS-SEM using confirmatory composite analysis. *Journal of Business Research, 109*, 101–110.

Hair, J. F., Hult, G. T. M., Ringle, C. M., & Sarstedt, M. (2017). *A primer on partial least squares structural equation modeling (PLS-SEM)*. SAGE.

Hair, J. F., Hult, G. T. M., Ringle, C. M., Sarstedt, M., Danks, N. P., Ray, S., et al. (2021). Evaluation of reflective measurement models. In *Partial least squares structural equation modeling (PLS-SEM) using R: A workbook* (pp. 75–90). Springer.

Hair, J. F., Ringle, C. M., & Sarstedt, M. (2011). PLS-SEM: Indeed a silver bullet. *Journal of Marketing theory and Practice, 19*(2), 139–152.

Hair, J. F., Sarstedt, M., Hopkins, L., Kuppelwieser, G., & V. (2014). Partial least squares structural equation modeling (PLS-SEM) An emerging tool in business research. *European Business Review, 26*(2), 106–121.

Han, T. I., & Park-Poaps, H. (2024). Egoistic and altruistic claims in organic cotton apparel advertising. *Journal of Global Fashion Marketing, 15*(2), 235–252.

He, H., Li, Y., & Harris, L. (2012). Social identity perspective on brand loyalty. *Journal of Business Research, 65*(5), 648–657.

Heim, H. (2022). Change of mind: Marketing social justice to the fashion consumer. *International Journal for Crime, Justice and Social Democracy, 11*(2), 102–115.

Henninger, C. E., Alevizou, P. J., & Oates, C. J. (2016). What is sustainable fashion? *Journal of Fashion Marketing and Management: An International Journal, 20*(4), 400–416.

Henseler, J., Hubona, G., & Ray, P. A. (2016). Using PLS path modeling in new technology research: Updated guidelines. *Industrial Management & Data Systems, 116*(1), 2–20.

Henseler, J., Ringle, C. M., & Sarstedt, M. (2015). A new criterion for assessing discriminant validity in variance-based structural equation modeling. *Journal of the Academy of Marketing Science, 43*, 115–135.

Heo, J., & Muralidharan, S. (2019). What triggers young Millennials to purchase eco-friendly products?: The interrelationships among knowledge, perceived consumer effectiveness, and environmental concern. *Journal of Marketing Communications, 25*(4), 421–437.

Hill, J., & Lee, H. H. (2012). Young Generation Y consumers' perceptions of sustainability in the apparel industry. *Journal of Fashion Marketing and Management: An International Journal, 16*(4), 477–491.

Ho, T. T. H., Vu, T. N. P., & Vu, H. M. (2020). Determinants influencing consumers purchasing intention for sustainable fashion: Evidence from Ho Chi Minh City. *The Journal of Asian Finance, Economics and Business, 7*(11), 977–986.

Hong, Y., Al Mamun, A., Masukujjaman, M., & Yang, Q. (2024). Sustainable consumption practices among Chinese youth. *Humanities and Social Sciences Communications, 11*(1), 1–17.

https://about.underarmour.com/en/Purpose/Sustainability/environment.html/ (accessed 24 Aug 2024a).

https://consciousfashion.co/guides/sustainable-fashion-marketing (accessed 24 Aug 2024b).

https://corporate.marksandspencer.com/sustainability/ (accessed 12 July 2024c).

https://earth.org/fast-fashions-detrimental-effect-on-the-environment/ (accessed 12 July 2024d).

https://fashionunited.com/global-fashion-industry-statistics/ (accessed 06 Sept 2024e).

https://global.diesel.com/ (accessed 21 Oct 2024f).

https://global.diesel.com/for-responsible-living/ (accessed 02 Oct 2024g).

https://keyhole.co/blog/nike-social-media-strategy/ (accessed 24 Sept 2024h).

https://news.adidas.com/ (accessed 24 Sept 2024i).

https://nuvonda.ch/en/pages/ecovero/ (accessed 31 Mar 2025a).

https://sustainablebrands.com/read/press-release/why-social-media-and-sustainability-should-go-hand-in-hand; https://www.instagram.com/reel/BEVRSjDgURd/ (accessed 12 Dec 2024j).

https://www.cognitivemarketresearch.com/organic-cotton-market-report/ (accessed 06 Sept 2024k).

https://www.ecovero.com/ (accessed 31 Mar 2025b).

https://www.eea.europa.eu/publications/environment-and-climate-pressures-from/ (accessed 31 Mar 2025c)

https://www.hm.com/om/lee-x-hm/ (accessed 06 Sept 2024l).

https://www.instagram.com/adolfodominguezofficial/p/CFoicmDi-L5/ (accessed 31 Mar 2025d).

https://www.invisionapp.com/inside-design/adidas-films/ (accessed 31 Mar 2025e).

https://www.levi.com/ (accessed 17 Sept 2024m).

https://www.levistrauss.com/2021/04/22/levis-launches-buy-better-wear-longer-campaign/ (accessed 25 Dec 2024n).

https://www.linkedin.com/pulse/wool-fabric-apparel-market-research-report-unlocks-analysis-wnt4e/ (accessed 06 Nov 2024o).

https://www.longchamp.com/us/en/maison-longchamp/sustainable/recycled-materials.html/ (accessed 06 Nov 2024p).

https://www.mavi.com/mavi-hemp/ (accessed 10 Aug 2024q).

https://www.nike.com/sustainability/ (accessed 10 Aug 2024r).

https://www.nike.com/sustainability/materials/ (accessed 10 Aug 2024s).

https://www.tentree.com/pages/sustainability/ (accessed 26 Feb 2025f).

https://www.wrangler.com/ (accessed 10Aug 2024t).

Hu, L. T., & Bentler, P. M. (1999). Cutoff criteria for fit indexes in covariance structure analysis: Conventional criteria versus new alternatives. *Structural Equation Modeling: A Multidisciplinary Journal, 6*(1), 1–55.

Hur, E., & Cassidy, T. (2019). Perceptions and attitudes towards sustainable fashion design: Challenges and opportunities for implementing sustainability in fashion. *International Journal of Fashion Design, Technology and Education, 12*(2), 208–217.

Iran, S. (2018). *Collaborative fashion consumption: A viable innovative concept of sustainable fashion consumption?* Doctoral dissertation, Technische Universität Berlin.

Isabella, G. (2022). Persuasion and the six Cialdini principles applied in social marketing. In *The Palgrave encyclopedia of social marketing* (pp. 1–8). Springer International Publishing.

Jackowski, T., Chylewska, B., Cyniak, D., Czekalski, J., & Jackowska-Strumiłło, L. (2003). Modelling of the relationship between feeding sliver structures and parameters of cotton/linen blended yarns. *Fibres & Textiles in Eastern Europe, 11*(2 (41)), 12–17.

Jackson, T. (2004). Negotiating Sustainable Consumption: A review of the consumption debate and its policy implications. *Energy & Environment, 15*(6), 1027–1051.

Jestratijevic, I., & Rudd, N. A. (2018). Six forms of sustainable fashion. *Latest Trends in Textile and Fashion Designing, 2*(4), 220–222.

Jiang, S., Jotikasthira, N., & Pu, R. (2022). Toward sustainable consumption behavior in online education industry: The role of consumer value and social identity. *Frontiers in Psychology, 13*, 865149.

Joy, A., Sherry, J. F., Jr., Venkatesh, A., Wang, J., & Chan, R. (2012). Fast fashion, sustainability, and the ethical appeal of luxury brands. *Fashion Theory, 16*(3), 273–295.

Judge, M., Warren-Myers, G., & Paladino, A. (2019). Using the theory of planned behavior to predict intentions to purchase sustainable housing. *Journal of Cleaner Production, 215*, 259–267.

Jung, J., Kim, S. J., & Kim, K. H. (2020). Sustainable marketing activities of traditional fashion market and brand loyalty. *Journal of Business Research, 120*, 294–301.

Khan, O., Varaksina, N., & Hinterhuber, A. (2024). The influence of cultural differences on consumers' willingness to pay more for sustainable fashion. *Journal of Cleaner Production*, 141024.

Kim, W., Ryoo, Y., Yoon, S., & Kim, K. (2021). Ethical dissonance in environmental advertising: Moderating effects of self-benefit versus other-benefit appeals. *International Journal of Advertising, 40*(8), 1320–1342.

Kotler, P. (1970). Corporate models: Better marketing plans. *Harvard Business Review, 48*, 135–149.

Kotler, P., & Lee, N. (2008). *Social marketing: Influencing behaviors for good.* Sage.

Kozlowski, R., & Muzyczek, M. (2023). Hemp, flax and other plant fibres. In *Sustainable Fibres for Fashion and Textile Manufacturing* (pp. 75–93). Woodhead Publishing.

Lambert, E. (2019). *Sustainable fashion consumption: Theory of planned behavior and the influence of self-identity, perceived consumer effectiveness and fashion consciousness.*

Lawson, L., Degenstein, L. M., Bates, B., Chute, W., King, D., & Dolez, P. I. (2022). Cellulose textiles from hemp biomass: Opportunities and challenges. *Sustainability, 14*(22), 15337.

Lee, C. K. C., Yap, C. S. F., & Levy, D. S. (2016). Place identity and sustainable consumption: Implications for social marketing. *Journal of Strategic Marketing, 24*(7), 578–593.

Legere, A., & Kang, J. (2020). The role of self-concept in shaping sustainable consumption: A model of slow fashion. *Journal of Cleaner Production, 258*, 120699.

Leyva-Hernández, S. N., Terán-Bustamante, A., & Martínez-Velasco, A. (2023). COVID-19, social identity, and socially responsible food consumption between generations. *Frontiers in Psychology, 14*, 1080097.

Lim, W. M. (2017). Inside the sustainable consumption theoretical toolbox: Critical concepts for sustainability, consumption and marketing. *Journal of Business Research, 78*, 69–80.

Lin, Y. T., Tseng, T. H., Chang, A., & Yang, C. C. (2022). A value adoption approach to sustainable consumption in retail stores. *International Journal of Retail & Distribution Management, 50*(11), 1412–1435.

Lunden, S., Sundström, L., & Suliman, A. (2020). *The effect of marketing appeals on consumers' intention to pro-environmental behavior: A social marketing study applying the Theory of planned behavior in Jönköping, Sweden.*

Maichum, K., Parichatnon, S., & Peng, K. C. (2016). Application of the extended theory of planned behavior model to investigate purchase intention of green products among Thai consumers. *Sustainability, 8*(10), 1077.

Majumdar, A., Shukla, S., Singh, A. A., & Arora, S. (2020). Circular fashion: Properties of fabrics made from mechanically recycled poly-ethylene terephthalate (PET) bottles. *Resources, Conservation and Recycling, 161*, 104915.

Mansour, O. S. M. (2017). Integrating the concept of modular design and dart manipulation technique for the innovation of fashion design for women. *International Design Journal, 7*(4), 87–97.
McKinney, E., Cho, S., Zhang, L., Eike, R., & Sanders, E. (2020). Analysis of zero waste patternmaking approaches for application to apparel. In *Sustainability in the textile and apparel industries: Sustainable textiles, clothing design and repurposing* (pp. 31–51). Springer.
Min, Y. W., Kim, J., & Jin, B. E. (2025). Are you truly buying sustainable clothing for the environment? Effective benefit appeals by item types and their link to public self-consciousness. In *International textile and apparel association annual conference*.
Mishra, S., Shukla, Y., Malhotra, G., Chatterjee, R., & Rana, J. (2023). Millennials' self-identity and intention to purchase sustainable products. *Australasian Marketing Journal, 31*(3), 199–210.
Mo, Z., Liu, M. T., & Liu, Y. (2018). Effects of functional green advertising on self and others. *Psychology & Marketing, 35*(5), 368–382.
Mukattash, I. L., Dandis, A. O., Thomas, R., Nusair, M. B., & Mukattash, T. L. (2023). Social marketing, shock advertising and risky consumption behavior. *International Journal of Emerging Markets, 18*(8), 1994–2011.
Mukendi, A., Davies, I., Glozer, S., & McDonagh, P. (2020). Sustainable fashion: Current and future research directions. *European Journal of Marketing, 54*(11), 2873–2909.
Nayak, R. (Ed.). (2022). *Sustainable fibres for fashion and textile manufacturing*. Woodhead Publishing.
Niinimäki, K. (2010). Eco-clothing, consumer identity and ideology. *Sustainable Development, 18*(3), 150–162.
Noble, G., Pomering, A., Johnson, W., & L. (2014). Gender and message appeal: Their influence in a pro-environmental social advertising context. *Journal of Social Marketing, 4*(1), 4–21.
Overcash, M. (2012). A comparison of reusable and disposable perioperative textiles: Sustainability state-of-the-art 2012. *Anesthesia & Analgesia, 114*(5), 1055–1066.
Papadas, K. K., Avlonitis, G. J., & Carrigan, M. (2017). Green marketing orientation: Conceptualization, scale development and validation. *Journal of Business Research, 80*, 236–246.
Patel, J. D., Trivedi, R. H., & Yagnik, A. (2020). Self-identity and internal environmental locus of control: Comparing their influences on green purchase intentions in high-context versus low-context cultures. *Journal of Retailing and Consumer Services, 53*(102), 003.
Pérez, A., Collado, J., & Liu, M. T. (2022). Social and environmental concerns within ethical fashion: General consumer cognitions, attitudes and behaviors. *Journal of Fashion Marketing and Management: An International Journal, 26*(5), 792–812.
Periyasamy, A. P., & Militky, J. (2020). Sustainability in regenerated textile fibers. In *Sustainability in the textile and apparel industries: Sourcing synthetic and novel alternative raw materials* (pp. 63–95). Springer.
Pinho, M., & Gomes, S. (2023). What role does sustainable behavior and environmental awareness from civil society play in the planet's sustainable transition. *Resources, 12*(3), 42.
Polonsky, M. J. (2011). Transformative green marketing: Impediments and opportunities. *Journal of Business Research, 64*(12), 1311–1319.
Pounders, K., Lee, S., & Royne, M. (2018). The effectiveness of guilt and shame ad appeals in social marketing: The role of regulatory focus. *Journal of Current Issues & Research in Advertising, 39*(1), 37–51.
Prakash, C. (2020). Bamboo fibre. In *Handbook of natural fibres* (pp. 219–229). Woodhead Publishing.
Pura, M. (2005). Linking perceived value and loyalty in location-based mobile services. *Managing Service Quality: An International Journal, 15*(6), 509–538.
Quoquab, F., Mohamed Sadom, N. Z., & Mohammad, J. (2021). Sustainable marketing. In *The Palgrave handbook of corporate social responsibility* (pp. 1269–1292). Springer International Publishing.
Ramjaun, T. A. (2021). Exploring the# zerowaste lifestyle trend on Instagram. In *The sustainability debate: Policies, gender and the media* (Critical studies on corporate responsibility, governance and sustainability) (Vol. 14, pp. 205–220). Emerald Publishing Limited.

Rask, A. (2021). *"Attitude is a little thing that makes a big difference": Exploring the impact of marketing appeals on sustainable consumption.*

Ruan, T., Sang, D., & Zeng, Q. (2022). Research on ZARA's business environment and marketing strategies in China. In *2022 2nd International Conference on Enterprise Management and Economic Development (ICEMED 2022)* (pp. 659–672). Atlantis Press.

Ruiter, R. A., Abraham, C., & Kok, G. (2001). Scary warnings and rational precautions: A review of the psychology of fear appeals. *Psychology and health, 16*(6), 613–630.

Ryoo, Y., Sung, Y., & Chechelnytska, I. (2020). What makes materialistic consumers more ethical? Self-benefit vs. other-benefit appeals. *Journal of Business Research, 110*, 173–183.

Salazar, H. A., Oerlemans, L., & van Stroe-Biezen, S. (2013). Social influence on sustainable consumption: Evidence from a behavioral experiment. *International Journal of Consumer Studies, 37*(2), 172–180.

Schutter, M. (2020). *A new formula for sustainable clothing consumption: The effects of regulatory focus framing and guilt appeals on intention to reduce clothing consumption.* Bachelor's thesis, University of Twente.

Seibel, S., Santos, I. H., & Silveira, I. (2021). Covid-19's impact on society, fashion trends and consumption. *Strategic design research journal, 14*(1).

Sermboonsang, R., Tansuhaj, P. S., Silpakit, C., & Chaisuwan, C. (2020). Mindfulness-based transformational learning for managing impulse buying. *Journal of Education for Business, 95*(2), 129–137.

Seva, R. R. (2022). Hidden power of affective products and environments. *Work, 73*(s1), S5–S14.

Shan, M., Zhu, Z., Song, C., & Chen, H. A. (2023). The effectiveness of advertising appeals: A culturally-derived power perspective. *Journal of Retailing and Consumer Services, 75*(103), 522.

Sharma, N., Saha, R., Sreedharan, V. R., & Paul, J. (2020). Relating the role of green self-concepts and identity on green purchasing behavior: An empirical analysis. *Business Strategy and the Environment, 29*(8), 3203–3219.

Sheth, J. N., & Parvatiyar, A. (2013). Ecological imperatives and the role of marketing. In *Environmental marketing* (pp. 3–20). Routledge.

Sheth, J. N., Sethia, N. K., & Srinivas, S. (2011). Mindful consumption: A customer-centric approach to sustainability. *Journal of the academy of marketing science, 39*, 21–39.

Stanković, S. B., Novaković, M., Popović, D. M., Poparić, G. B., & Bizjak, M. (2019). Novel engineering approach to optimization of thermal comfort properties of hemp containing textiles. *The Journal of The Textile Institute, 110*(9), 1271–1279.

Stephen, J. (2023). *Broadening the path to self-actualization through systematic changes to Maslow's hierarchy of needs.* Doctoral dissertation, Baylor University.

Stets, J. E., & Burke, P. J. (2000). Identity theory and social identity theory. *Social Psychology Quarterly*, 224–237.

Storey, J. D., Hess, R., & Saffitz, G. (2015). Social marketing. In K. Glanz, B. K. Rimer, & K. Viswanath (Eds.), *Health behavior: Theory, research, and practice* (5th ed., pp. 411–438). Wiley.

Tajfel, H. (1979). Individuals and groups in social psychology. *British Journal of social and clinical psychology, 18*(2), 183–190.

Temel, A. (2023). An integration of self-conscious emotion into public service announcement: The effect of framing messages based on guilt and shame-inducing strategies on the intention of behavior change. *Mavi Atlas, 11*(2), 422–441.

Teng, C. C., Lu, A. C. C., & Huang, T. T. (2018). Drivers of consumers' behavioral intention toward green hotels. *International Journal of Contemporary Hospitality Management, 30*(2), 1134–1151.

Tian, K. T., Bearden, W. O., & Hunter, G. L. (2001). Consumers' need for uniqueness: Scale development and validation. *Journal of Consumer Research, 28*, 50–66.

Türkmen, N. (2009). *A new design philosophy for sustainability: Slow design* (pp. 869–874). Autex.

Valtanen, L. (2021). *Consumer motivations and value creation from sustainable fashion consumption: A qualitative study of sustainable fashion consumers.* Bachelor's Thesis, Aalto University.

Van Breda, L., Terblanche-Smit, M., & Pelser, T. (2023). The effectiveness of sustainability social marketing use of fear and guilt appeals to influence the behavioral intention of millennials. *European Business Review, 35*(2), 202–222.

Van den Berg, J. (2020). *The effect of self-benefit and other-benefit appeals in combination with emotional message framing on the consumption of demerit goods.* Master Thesis, Erasmus University Rotterdam.

Van der Werff, E., Steg, L., & Keizer, K. (2013). The value of environmental self-identity: The relationship between biospheric values, environmental self-identity and environmental preferences, intentions and behavior. *Journal of Environmental Psychology, 34*, 55–63.

Varey, R. J. (2010). Marketing means and ends for a sustainable society: A welfare agenda for transformative change. *Journal of Macromarketing, 30*(2), 112–126.

Vesterinen, E., & Syrjälä, H. (2022). Sustainable anti-consumption of clothing: A systematic literature review. *Cleaner and Responsible Consumption, 5*(100), 061.

Vladimirova, K., Henninger, C. E., Alosaimi, S. I., Brydges, T., Choopani, H., Hanlon, M., et al. (2024). Exploring the influence of social media on sustainable fashion consumption: A systematic literature review and future research agenda. *Journal of Global Fashion Marketing, 15*(2), 181–202.

Voorhees, C. M., Brady, M. K., Calantone, R., & Ramirez, E. (2016). Discriminant validity testing in marketing: An analysis, causes for concern, and proposed remedies. *Journal of the academy of marketing science, 44*, 119–134.

Wang, Z., Tao, X., Zeng, X., Xing, Y., Xu, Z., & Bruniaux, P. (2023). Design of customized garments towards sustainable fashion using 3D digital simulation and machine learning-supported human–product interactions. *International Journal of Computational Intelligence Systems, 16*(1), 16.

Wei, X., & Jung, S. (2017). Understanding Chinese consumers' intention to purchase sustainable fashion products: The moderating role of face-saving orientation. *Sustainability, 9*(9), 1570.

Wei, X., & Jung, S. (2022). Benefit appeals and perceived corporate hypocrisy: Implications for the CSR performance of fast fashion brands. *Journal of Product & Brand Management, 31*(2), 206–217.

Weinreich, N. K. (2012). *What is social marketing? Weinreich Communications.* https://www.social-marketing.com/what-is-social-marketing (accessed 24 Jan 2025).

White, K., & Peloza, J. (2009). Self-benefit versus other-benefit marketing appeals: Their effectiveness in generating charitable support. *Journal of Marketing, 73*(4), 109–124.

Williams, K. C. (2012). Fear appeal theory. *Research in Business and Economics Journal, 5*(1), 1–21.

Xu, J. (2022). The impact of guilt and shame in charity advertising: The role of self-construal. *Journal of Philanthropy and Marketing, 27*(1), e1709.

Yang, D., Lu, Y., Zhu, W., & Su, C. (2015). Going green: How different advertising appeals impact green consumption behavior. *Journal of Business Research, 68*(12), 2663–2675.

Yang, S., Song, Y., & Tong, S. (2017). Sustainable retailing in the fashion industry: A systematic literature review. *Sustainability, 9*(7), 1266.

Yang, Y., Hao, X. M., Hao, X., Li, H. W., & Feng, X. X. (2013). Investigate on moisture absorption/desorption of shoe material containing hemp fiber. *Advanced Materials Research, 627*, 49–52.

Yousef, M., Dietrich, T., & Torrisi, G. (2021). Positive, negative or both? Assessing emotional appeals effectiveness in anti-drink driving advertisements. *Social Marketing Quarterly, 27*(3), 195–212.

Zhang, R. (2014). *Sustainable apparel consumption: Scale development and validation.* Doctoral Dissertation, Oregon State University.

Zollo, L. (2021). The consumers' emotional dog learns to persuade its rational tail: Toward a social intuitionist framework of ethical consumption. *Journal of Business Ethics, 168*(2), 295–313.

Zott, C. (2024). *Ecoalf's Inspiring Business Model: Turning Ocean Trash into Sustainable Fashion.* https://www.linkedin.com/pulse/ecoalfs-inspiring-business-model-turning-ocean-trash-sustainable-5or2f/ (accessed 18 Feb 2025).

Modern Tools in Fashion Retailing and Consumer Perspective

Keywords Modern tecnology · Modern tools · Smart retailing · Service quality · Mass customization · Experiential marketing · Consumer behavior · Unified Theory of Acceptance and Use of Technology (UTAUT) models · Stimulus-Organism-Behavior (SOR) model

1 Introduction

The emergence of data science technologies reshaped every business and ensured a digital transformation. The retailing industry is no exception. Whereas the supply chain operations improved with the optimization algorithms, the use of communication and interaction technologies in shopping environments resulted in smart retailing. Smart retailing is defined as "*an interactive and connected retail system, which supports the seamless management of different customer touchpoints to personalize the customer experience across different touchpoints and optimize performance over these touchpoints*" (Roy et al., 2017), and smart retailers utilize the data science technologies and various tools working on them, such as smart shelves, interactive displays, smart fitting rooms, shopping assistance systems, touchless payment systems, smart shopping carts, and self-checkout systems to increase their competitiveness.

Retailers have adopted modern technology and tools to improve their value chain operations, enhance service quality, deliver customized products and services and provide an improved customer experience. The potential economic impact of smart technology is expected to rise from $410 billion to $1.2 trillion by 2025 in retail settings (Ng et al., 2021). Nonetheless, the relevant technology and the pace and ways of adoption are different for retailers specialized in various product categories. On the other hand, the implementation of smart retail technology provided consumers with greater convenience, accessibility, flexibility, and efficiency, as well as greater engagement and joyful experiences. However, consumers have various

C. Saricam, N. Okur, *Changing Textile and Apparel Consumption in Transformative Era of Sustainability and Digitalization*, Sustainable Textiles: Production, Processing, Manufacturing & Chemistry,
https://doi.org/10.1007/978-3-031-97419-9_5

expectations from smart tools and therefore show different tendencies for them. Besides, they approach smart tools and technology with different motivations. Therefore, it is important to assess smart retailing based on specific retail product categories and from the consumers' perspective.

Considering these points, this chapter aims to explain the modern tools and technology in the transformative era with a specific focus on fashion retailing. In this respect, the chapter begins with general information about modern technologies, including some facts showing their adoption and implementation rate. The chapter continues with explanations of modern tools from retailers' and consumers' perspectives. Specifically, the modern tools working on modern technology are introduced with specific examples and cases from fashion retailers. Then the consumers' approach towards smart retailing is explained with a discussion on certain motivations, expectations, and consumer characteristics. The chapter ends with an empirical study, which concentrates on analyzing consumer behavior towards modern tools in mobile fashion retailing with an additional discussion on consumer characteristics.

2 Modern Technologies and Their Use in Retailing

This section defines the main characteristics of emerging technologies, followed by their possible use in retailing activities. Artificial intelligence, the Internet of Things, Big Data, blockchain, and robotics are introduced in the first subsection, whereas augmented reality, virtual reality, and virtual try-on technologies will be explained in the second subsection, as they have special significance for the fashion retail industry.

2.1 Types of Modern Technology in Smart Stores

In this subsection, information will be given about Artificial Intelligence, Big Data, the Internet of Things, blockchain, RFID, QR codes, and robotics. There is a huge amount of investment in these technologies because of their possible contribution to industries and the economy. Artificial intelligence and many relevant emerging technologies are expected to contribute \$15.7 trillion to the world economy by 2030 (Oliveira, 2019). Specifically, the market of AI technologies is calculated to be around 244 billion U.S. dollars in 2025 and is expected to be over 800 million U.S. dollars by 2030 (Thormundsson, 2025). Not only AI-based technologies but also other technologies such as Big Data, IoT, robotics, and blockchain gained importance. In this respect, global data creation is estimated to be more than 394 zettabytes (Statista, 2024a). The IoT market is expected to reach US\$1.06tn by 2025 worldwide (Statista, n.d.). The service robotics industry is projected to reach a market value of \$84.8 billion by 2028 (Arora et al., 2024). Blockchain solutions were estimated to reach almost 19 billion U.S. dollars by 2024 (Statista Research Department, 2025).

Artificial Intelligence (AI) AI is defined as "*computing systems that are able to engage in human-like processes such as learning, adapting, synthesizing, self-correction, and use of data for complex processing tasks*" (Crompton & Song, 2021). It can also be described as an activity that makes machines intelligent enough to function appropriately with foresight in an environment (Pillai et al., 2020). Being inspired basically by the human brain, AI is based on the concept of neural networks and computational models of these networks. It makes the devices sense, think, and act more independently while doing tasks (Soni, 2021).

Dating back to the 1950s, AI has many subcategories, such as machine learning, deep learning, logical reasoning and probabilistic reasoning, expert systems, computer vision, natural language processing, speech recognition, semantic understanding, and generative AI. Logical reasoning refers to drawing conclusions based on premises and evidence, whereas probabilistic reasoning refers to arriving at conclusions in the face of uncertainty. Expert systems mimic human intelligence by applying a set of rules and a knowledge base (Oliveira, 2019). Machine learning encompasses techniques that learn from data sets by recognizing the patterns and laws to make predictions afterwards (Weber & Schütte, 2019). Deep learning is a more specialized field in machine learning. The term "deep" in deep learning corresponds to the multiple layers in artificial neural networks (Soni, 2021). Computer vision technology, such as image recognition and facial recognition, involves the interpretation and understanding of visual information similar to how humans perceive it (Soni, 2021). Natural language processing technology investigates how computers can interpret and manipulate text and speech written and spoken in natural languages (Oliveira, 2019). Speech recognition and semantic understanding help machines to both listen and understand humans and even understand emotions (Zhang & Lu, 2021). On the other hand, generative AI is one type of intelligence that can create novel content and concepts such as narratives, pictures, videos, and melodies (Oliveira, 2019).

AI can be employed in the development of various tools and can effectively be utilized in both retail operations and customer services. In retail operations, AI might ensure automation in work processes that require routine and repeating tasks. It can optimize the processes in demand forecasting and manage the inventory levels efficiently with reduced operational costs by eliminating the shortage and overstock. Moreover, it is used effectively to build profitable pricing strategies with better analyses of the competition and market dynamics (Khan & Ahmed, 2024). Specifically in the fashion industry, AI is employed to make a detailed fashion style and trend analysis and even create fashion styles (Jain, & Wah, 2022). On the other hand, AI enables retailers to give better service at various touchpoints with customers. The store layout can be optimized, the product trials can be enhanced, and the payments can be facilitated for consumers. By using advanced algorithms and machine learning techniques, consumers' preferences, requirements, and consumption habits can be analyzed in detail and used to give personalized recommendations (Raji et al., 2024).

Recommendation systems constitute one specific field of AI-based applications. They use a combination of tools and technologies, which help consumers access the products they are seeking, obtain broader knowledge about the products, and reach personalized suggestions for making purchase decisions (Pereira et al., 2022). The recommendation systems have mainly two major parts: the customer model and the recommendation model. The information in the form of text and images for a recommendation system is collected either explicitly via consumer interactions such as rating items and ranking items or implicitly via consumer monitoring mechanisms such as customer views (and viewing times) of items in the store, purchase history, and social network analysis. Information extraction algorithms in consumer models extract high levels of structured data and customer profiles, which are dynamically updated over time. The data filtering tools, namely content-based filtering and collaborative filtering in the recommendation models, match product or service characteristics to a specific customer. The products are filtered based on the past behavior of consumers in content-based filtering, whereas they are filtered according to the preferences of similar consumers in collaborative filtering (Pereira et al., 2022).

Big Data Analytics Big Data refers to both the collection of massive amounts of data and the advanced digital technologies used to reveal patterns related to it (Favaretto et al., 2020). Big data is traditionally characterized by the term 5Vs, namely, volume, variety, velocity, veracity, and value. Whereas volume represents the amount of data generated from various resources, like social media, IoT devices, and also customer transactions, variety represents the number of formats the data can be found in. Velocity means the speed at which the data is gathered and processed. Veracity corresponds to the accuracy and trustworthiness of the data, and value refers to the potential of the data to generate actionable insights (Kannan et al., 2024). The main contribution of big data is that it can give insights from huge data sets. On the other hand, big data analytics refers to the advanced analytics methods that discover hidden patterns by using complex applications with elements such as predictive models, statistical algorithms, and what-if analysis powered by analytics systems (Kannan et al., 2024; Prasad & Venkatesham, 2021).

Big data is gathered from various resources in retailing, such as online transactions, customer accounts and loyalty programs, website and application analytics, customer surveys, store interactions, and numerous in-store devices providing information about heavy traffic and waiting times. Big data analytics can be used in the optimization of retail operations such as sales forecasting, trend forecasting, and inventory management together with AI. Thus, Big Data and Big Data Analytics can be used in analyzing and segmenting the market, making consumer behavior analyses, and providing better service to consumers. Specifically, Big Data makes it possible to identify the most valuable consumer groups and to get a deeper insight into an individual customer (Prasad & Venkatesham, 2021). The collection of data in various shapes and forms (a name, a photo, an email address, bank details, posts on social networking websites, medical information, or a computer IP address) enables the richness of consumer behavior analysis (Favaretto et al., 2020). The results can be used to optimize the product assortments and customize the products and services given to consumers (Fisher & Raman, 2018).

Internet of Things (IoT) Covering a broad class of connected devices that are remotely accessed through the internet, IoT is defined as "*the interconnected network of physical objects (devices, vehicles, equipment, buildings, etc.) that are embedded with sensors, software, and network connectivity, allowing them to collect and exchange*" (Jaiswal et al., 2023). The IoT ecosystem consists of various sensors and actuators, including the environmental, temperature, optical, and motion sensors for remote control and the wearable and implantable sensors for biological use, and various components, basically the network technologies, including Wi-Fi, Bluetooth, Bluetooth Low Energy (beacons), Zigbee and Z-Wave, and Low Power Wide Area Networks (Caro & Sadr, 2019). IoT can be used in both supply and value creation in the retailing industry with an integration of AI-based processes.

The primary benefit of IoT in supply chain and manufacturing is industrial automation. IoT ensures the collection of massive amounts of real-time data from a network of devices and systems. The interconnected devices exchange data and make decisions in real time. Specifically, sensors receive information from machinery and communicate this information to adjust machine operation and alarm the human operator. In this respect, IoT helps to automate complicated procedures and facilitates monitoring and maintenance by allowing businesses to identify problems before they occur. Moreover, IoT optimizes production processes by identifying bottlenecks and inefficiencies and improves product quality by ensuring that they meet standards and specifications (Jaiswal et al., 2023). In this way, it improves efficiency and decision-making, enhances safety, and reduces expenses.

IoT can also be efficiently used in logistics and retail operations. It can optimize the movement of goods and automate the planning of routes and scheduling in logistic operations. It helps better management of inventory and optimizes the supply chain operations on the supply side.

IoT enhances the consumer experience and enables personalized marketing on the retail side (Jaiswal et al., 2023). The sensors in devices such as smart shelves, camera systems in stores, beacons placed all over the store, barcodes, QR codes, RFID tags, and near-field communication readers can all provide data that can be analyzed to direct consumers in stores (Hossain et al., 2021). Moreover, the IoT can be used to communicate with consumers by creating an ecosystem whose main layer involves the IoT gadgets (smartphones, tablets, or other gadgets) and the social channels connected via the Internet (Hossain et al., 2021). The location-based services pinpoint the location of the customer and might present personalized notifications to consumers and improve the shopping experience.

Even though it has the potential to work individually, the IoT closely collaborates with Big Data and AI technologies. In this respect, IoT facilitates the formation of Big Data. Big Data applications store the data, and AI extracts meaningful insights from the data with machine learning and deep learning algorithms in return.

Radio Frequency Identification (RFID) and Quick Response Codes (QR Codes) Using a special tag or label embedded with a computer chip and an antenna, RFID is a technology that communicates with a reader or transponder (Koh et al., 2006). It can be considered a special field of IoT as it transmits and communicates data in IoT systems. RFID can successfully be used to improve operational efficiency and accuracy in inventory management systems. Specifically, it reduces shrinkage, improves inventory control, and reduces labor costs (Koh et al., 2006). Moreover, it helps retailers to provide features like item-level visibility, which can quickly locate the desired item and integrate with the other technologies in interactive fitting rooms (Mythili et al., 2024).

QR codes are not considered as special fields of IoT, but they are used as a connection point in an IoT ecosystem. A QR code is a two-dimensional barcode that contains greater amounts of information (textual information, numerical values, and binary data) in both vertical and horizontal directions. The information can be decoded through mobile devices such as smartphones due to the embedded camera and a code reading software installed. QR codes help consumers to find information about the products, their origins, or care instructions, as well as to make the payment easily (Albăstroiu & Felea, 2015). On the other hand, RFID and QR codes integrated with AI and IoT allow retailers to eliminate counterfeits due to their capacity to store data about the origins of products (Caro & Sadr, 2019).

Robotic Technology Service robots are defined as autonomous and adaptable interfaces used in manufacturing and retail operations to interact. The service robots or the interfaces can be both humanoid (having the morphology of a human) and non-humanoid (without human-like characteristics) systems (De Gauquier et al., 2020). Besides, some of them perform cognitive-analytical processes, whereas others perform social interactions with humans (Roozen et al., 2023). Robots are employed for automation in warehouses, delivery, inventory management, and customer service. Humanoid service robots can act as welcome assistants, information agents, sales assistants, or brand ambassadors (Fiestas Lopez Guido et al., 2024). The integration of AI-based technologies makes the robots more autonomous and smart with capabilities like understanding speeches, having computer vision, and learning new things.

Blockchain Blockchain is a decentralized digital transaction database that stores financial records, ledgers, and other digital information. It can be accessed, filtered, and controlled by different users. The computers in the blockchain networks act as the nodes in the blockchain where the information is collected in blocks. The blockchain is preferred by the companies as it allows making transactions at any point, and it eliminates fraud and false information as a transaction is verified before it is authorized and added to the ledger. Moreover, blockchain maintains transparency and reduces personnel requirements (Saxena & Sarkar, 2023; Medida, 2020).

The blockchain is used in retailing for performing both the supply chain and retail operations. In regard to supply chain operations, blockchain allows the tracking of products in the production stages and eases inventory management with the automatic updating of inventory levels (Medida, 2020). It also enables secure payments, helping to prevent potential disputes. Smart contracts provided by blockchain enable automatic payment capability when the required conditions are satisfied (Jabbar et al., 2021). On top of that, blockchain-based payment systems require low costs compared to traditional financial tools.

On the retail side, blockchain provides secure ways to manage customer data, which eliminates the privacy concerns of consumers. Moreover, it prevents product forgery and counterfeiting. Via blockchain, the brands might allocate every product a unique scannable code or mark, and the customers might reach the entire history of that product (Medida, 2020). Blockchain can also be used to build customer loyalty programs in the most secure and transparent way (Utz et al., 2023; Medida, 2020).

On the other hand, blockchain is also the main infrastructure technology for cryptocurrencies (such as Bitcoin, Ethereum, and etc.) and nonrefundable tokens (NFTs). The main feature of cryptocurrencies to prevent double-spending can be provided by blockchain since blockchain can embed the transaction into a block and transactions history in the blockchain (Xu et al., 2019). Blockchain enables the development, storage, and transfer of NFTs and certifies the authenticity of digital assets. NFTs are digital certificates of ownership, which are attached to digital content secured on a blockchain. They are unique, and they cannot be replaced with a copy but can be turned into tradable digital assets (Park et al., 2023). Blockchain technology enables copyright protection for NFTs and ensures transparent, verifiable transactions (Sung et al., 2023).

It should be added that blockchain, NFTs, and other technologies such as augmented reality, virtual reality and the Metaverse are all accepted as Web 3.0 topics (Park et al., 2023). Web 3.0 is the third generation of the internet that allows the generation of data by users and ensures communication of users at any time regardless of devices or networks (Hinmikaiye et al., 2022). Depending on the blockchain technology, Web 3.0 replaces the central data storage with distributed data storage and organizes information in new ways.

2.2 Virtual Tryon Technologies, Augmented Reality, Virtual Reality

The global virtual fitting room market size is expected to grow to almost 15 billion U.S. dollars by 2029 (Statista, 2024b). Virtual reality and augmented reality are claimed to be one of the top ten strategic technologies expected to evolve. The worldwide market for augmented reality is expected to be \$198 billion in 2025, and the combined market size of the augmented reality and virtual reality sector has been projected to exceed \$571 billion by the year 2025 (Oyman et al., 2022; de Amorim et al., 2022). Additionally, the metaverse market, which serves as a key platform for the integration of augmented reality and virtual reality, is projected to grow significantly, reaching \$1.3 trillion by 2030 (Balakrishnan et al., 2024).

Virtual Try-on Technologies Virtual try-on, virtual fit/size technology, or virtual fitting technologies are all defined as image interactivity technology, which allows consumers to see the overlay of products on the virtual models based on consumers' body measurements or make virtual product trial experiences to view apparel products on a simulation of their bodies in order to evaluate garment fit and appearance (Tawira & Ivanov, 2023; Lee & Xu, 2020). These solutions differ from each other in terms of how the physical body measurements are gathered and uploaded into the system (Lee & Xu, 2020). In some systems, the body measurements are uploaded by using a sophisticated 3D scanner (BodymetricsTM), whereas customers provide a few body measurements manually in 3D avatars (MyVirtualModelTM) and 3D customer models (FitiquetteTM) in others. It is possible to process the features of photo and body metrics together in a Photoaccurate 3D customer model (triMirrorTM) or to use a physical robotic mannequin utilizing shape-lifting techniques in robotic mannequin systems (EuvekaTM). Camera-based motion-capturing technologies are used in AR fitting rooms (FittingBoxTM), and a nonpersonalized model is provided in VR fitting rooms (AliBaba's Buy+TM) (Lee & Xu, 2020). Even though the most accurate avatar is created using 3D body scan data, body scans can have holes in specific parts of the body, such as the armpit and crotch. All virtual try-ons are equipped with 3D virtual garment simulation systems. However, there might be some challenges in the simulation of correct garment size, fabric textures and wrinkles, and the accurate interaction between garment and body and/or environments (Kim & LaBat, 2013). Moreover, the avatars created with consumers' selected key body measurements or created with selected body types such as pear, apple, or rectangular shapes are never exact replicas (Tawira & Ivanov, 2023). The limitations to 3D virtual simulation technology are improved in the systems employing augmented and virtual reality technologies.

Moreover, it is possible to use AI in virtual try-ons. For instance, Forma and Zeekit are using AI-powered image processing, which allows users to create their own photorealistic avatars with the camera or make a selection of pre-loaded avatars with different body types (Tawira & Ivanov, 2023).

Augmented Reality (AR) Augmented reality (AR) is the practice of augmenting a real-time, direct, or indirect view of the physical world with virtual information (Scholz & Smith, 2016). In AR applications, the environment can be enriched with computer-generated images, words, and sounds and annotation (Hariharan et al. 2020). This means that the real imagery/environment and virtual 3D digital graphics/objects coexist in the same space (Kang et al., 2023). The three features of AR were listed as combining reality and virtuality, creating real-time interactions, and 3D display of the products (Oyman et al., 2022).

The conceptualization of AR goes back to the 1960s with the creation of the multisensory simulator "Sensorama." 3D graphics were developed by the research institutes in AR-enabled wearable technologies by NASA in the 1970s and 80 s. The term AR was used for the first time by Scientists Caudell and Mizell in the 1990s

(Oyman et al., 2022). The breakthrough in AR-based technologies occurred with the transfer of AR to mobile applications as a result of the widespread use of smartphones.

There are different types of AR applications. The AR-enabled active print/packaging is one type of AR application that has the capability to augment targets, which are presented in a magazine, on a product package, in a catalogue, or in out-of-home advertisements with digital objects, by using privately owned devices. One example of this is the IKEA catalogue, which presents the furniture in a user's room. Bogus windows are the second type of AR application. They enhance the user's view by augmenting the space with digital objects. The Pepsi Max bus shelter, which augments a live street scene with fantastic images such as UFOs and tigers, is an example of this type. The third type of AR application is named as Geo layer type of AR application, in which the space around users is augmented with digital objects that may or may not be linked to specific geolocations by using privately owned devices. The digital penguins in the Tokyo Aquarium guide are an example of this type of AR application. The final type of AR application is the magic mirrors that augment the space or objects around the user with digital objects via public devices such as TV screens and digital mirrors. In the magic mirror, the user sees himself as part of the augmentation. With these types of applications, users can virtually try on clothing and shoes using digital screens or in shop windows, as in the case of Timberland's mall campaign (Scholz & Smith, 2016).

AR has been used in various fields such as education, engineering, gaming, healthcare, the military, real estate, travel/tourism, and retailing, especially in the furniture, beauty cosmetics, and apparel categories (Hariharan et al., 2020). In retailing, AR-based tools can be implemented in web-based and mobile-based applications (Plotkina et al., 2022). In this way, the consumers might bring products home before making the real purchase and experience products comfortably in their homes. Besides, AR is widely used in product trials and virtual try-ons. The virtual try-ons allow consumers to evaluate the performance of beauty or fashion products on the face or body (Jiang et al., 2023; Dacko, 2017; Soon et al., 2020). Moreover, AR can be successfully used for advertising in marketing campaigns. An immersive brand narrative can be created, and meaningful consumer–brand relationships can be developed with AR technology (Scholz & Smith, 2016; Kang et al., 2023).

Virtual Reality (VR) Virtual Reality (VR) is defined as "*the environment created by a computer or other media, an environment in which the user feels present*" (Kim et al., 2021). Being a type of digital and immersive media, VR generates a three-dimensional, virtual, imaginary, and interactive environment (Herz & Rahe, 2020). Unlike AR, the simulated environment replaces the real environment in VR (Kim et al., 2021). VR is superior to other digital media because it presents a more impactful and enduring experience. It presents simulated multisensory (vision, touch, hearing) experiences and interactive processes, which allow dynamic product viewing, trial, and usage. The product experience is spatialized and contextualized. Moreover, social interactions are possible with virtual products and services by means of avatars or telepresence (Wedel et al., 2020).

As it is in AR, the simulator Sensorama, which lacked individual interaction at that time, is accepted as the starting point for VR. Improving the technology with the addition of interactive graphics, head tracking, and new image processing techniques allowed interaction with individuals (Farah et al., 2019).

VR technology basically brings many components together in order to introduce the user to the virtual world, such as the graphics system, tracking system, information system, and visual display systems (Vrechopoulos et al., 2009). The graphics system generates images with 3D simulations and rendering technology, which makes the virtual space more dynamic and interactive. The tracking system, integrated with motion sensors and cameras, reports and reflects the position and orientation of the user. The information system stores the features of the virtual space, objects, and spatial data. The visual display systems, head-mounted displays (HMDs) in general, allow the projected image to be seen through their built-in screens. HMD makes 3D places and characters appear life-sized without the boundaries of television or computer monitors. Coupled with a gyroscope, the head movements of the users can be sensed, and the display can be corrected accordingly (Sung, 2020).

The subjective and physiological responses evoking real environments are associated with the immersive capability of VR (Wedel et al., 2020). The immersive capability can be increased by three mechanisms: the illusion of depth, synthesized sound, and interactive devices. The illusion of depth is created by sending slightly different images to the eyes of users with stereoscopic imaging and used to improve the telepresence. The synthesized sound, including ambient sounds and audio sensations, is utilized to engage users with a virtual space (Sung, 2020). The interactive devices, such as tracked gloves with pushbuttons, substitute the real interactions in the physical world (Vrechopoulos et al., 2009).

The VR systems are usually classified according to immersive characteristics. VR systems utilizing desktop screens are usually identified as non-immersive VR, whereas VR systems employing head-mounted display (HMD) systems are identified as immersive VR (Kim et al., 2021). Nonetheless, it is possible to classify VR systems into three categories: non-immersive VR (desktop-based environments), semi-immersive VR (cave automatic virtual environments), and fully immersive VR (wearing a head-mounted display). Non-immersive VR, or desktop-based VR, is the most common, simplest, and easiest way of accessing VR applications, as they do not require users to wear equipment (Kim & Choo, 2023). Desktop-based VR enables users to interact with virtual products by using a flat screen, keyboard, or computer mouse. The virtual objects could be rotated, picked up, and watched from different angles by using two handheld controllers (van Berlo et al., 2021). In semi-immersive VR, or cave automatic virtual environments (CAVE), the virtual world is projected on the walls, ceiling, and floor of a room by diverse stereoscopic projectors. In the application of CAVE, the user must wear passive stereo glasses to achieve a 3D view of the virtual world. (Violante et al., 2019). In immersive VR, users wear head-mounted displays (HMD), which consist of active glasses with a small screen placed properly in front of each eye. The users' sensorimotor illusion is maximized by providing realistic stimulation to multiple sensory channels simultaneously (Kim & Choo, 2023).

VR can be used either in functional activities such as learning, training, prototyping, and shopping or in hedonic activities such as gaming, entertainment, or tourism (Herz & Rahe, 2020; Wedel et al., 2020). Nonetheless, VR is widely used in retailing and advertising. VR presents an environment closer to a physical store than an e-commerce website, which allows consumers to visually evaluate products in a 3D naturalistic environment and gain in-depth information on digital shopping (Kim et al., 2021). VR technologies can also be used in virtual try-ons, as in the case of AR. Moreover, VR is preferred in interactive advertising, which can increase consumer engagement. VR technologies are employed in virtual store visits, virtual fashion shows, and metaverse platforms. Specifically, marketers contact consumers via digital stores, digital or phygital products, and entertainment such as fashion shows or branded VR games by developing editorial content (Sung, 2020).

As implied previously, AR is much favored in the virtual try-ons, whereas VR is used in the metaverse applications. Nonetheless, there is also a possibility to use both AR and VR at the same time. In fact, the combination of virtual and augmented reality is called Mixed Reality (MR) or Extended Reality (XR). Extended reality includes the entire spectrum from "the complete real" to "the complete virtual" in the concept of reality-virtuality continuum (Hariharan et al. 2020). This enhancement of reality requires the use of different devices, such as handheld displays (e.g., smartphones, tablets), head-mounted displays (HMD, e.g., smart glasses), or special displays (Teles Roxo & Quelhas Brito, 2020). FaceCake, a recognized leader in personalized AR try-on, and Fitnect, augmented reality fitting room platforms for various clothes and accessories, use technically mixed-reality platforms, which can extend the functionality of their augmented reality platforms along with more immersive virtual environments (McQueen & Hamza-Lup, 2018).

3 Modern Tools and Technology within Retailing Perspectives

Data science and relevant technologies take place at the core of all retail activities, as shown in Fig. 1. They can be either directly used to provide solutions or indirectly used in the implementation of modern tools at the touchpoints with consumers. Retailers benefit from these technologies and tools to improve the proficiency of supply chain and retail operations, enhance the service quality for consumers, customize the products and services according to the preferences of their consumers, and present a memorable and joyful user experience. This section will describe the benefits of these tools and technologies by giving examples and cases from fashion brands under the four subsections below.

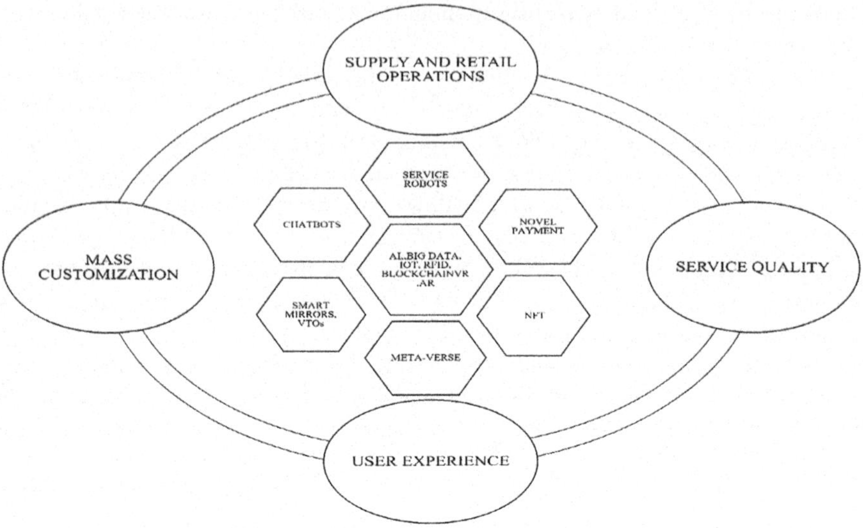

Fig. 1 The use of modern technology and tools by retailers

3.1 Use of Modern Tools as a Part of Retail Operations

The retail industry is the last link in the textile and fashion value chain. The retail operations involve many supply chain management activities, logistics and distribution activities, and marketing management. Improvements in these activities result in efficiency increases, cost reduction, and customer satisfaction. Modern technology having the potential to offer excellence and proficiency is particularly employed in production management, automation of production and logistics, inventory management, trend forecasting, product development, creative processes, product authentication, and enhancement of sustainability. Since the application areas cover many issues, all types of modern technologies find an application area. Nonetheless, AI-based algorithms, big data, IoT, blockchain, and robotics are the most convenient technologies for achieving proficiency in supply and retail operations.

The production operations need improvements in terms of flexibility, productivity, and better use of resources. Modern technology provides scalability and ensures faster production by enabling traceability and control without creating problems in quality. Specifically, scalability and quality production are provided with the employment of service robots in some operations. In this regard, Nike uses robotics to clean and repair the used Nike shoes in the Bot Initiated Longevity Lab (Öztürk, 2023). The most convenient solutions for faster production quality and control are possible by using data-driven, AI-based processes. In this regard, many production companies have adopted digital transformation. Alternatively, the fashion brands upgrade their systems to provide production management solutions for their value chain. Levi Strauss & Co. optimizes production processes by improving production schedules and decreasing the lead times based on data received from suppliers, manufacturers, and retailers. The optimization processes are extended to solving the problem at a bottleneck station in manufacturing units (Olabiyi, 2023).

Retail operations rely heavily on logistics operations for supply, warehousing, distribution, and, recently, reverse logistics. Better management of these activities ensures order fulfillment and on-time distribution and reduces costs. Service robots are usually employed in supply chain operations such as production and logistics. In addition to the movement of heavy packages, advanced robotics, equipped with artificial intelligence and computer vision, sorts packages efficiently by reducing in time requirement (Staff, 2022). Since 2012, Amazon has made investments in many types of robotics to increase productivity in supply chain activities.

Inventory management, as one of the core tasks of retailing, aims to decrease the stock levels and prevent the stock. It is facilitated with the use of data-driven tools and AI-based optimization. According to the report prepared by McKinsey, data-driven planning leads to a 5–15% reduction in inventory and a 15–25% reduction in stock-outs. Therefore, 75% of fashion executives prioritize using data-driven planning and forecasting tools, implying that 20% of profit is lost due to inaccurate stock buying (Balchandani et al., 2024). Many companies, such as Adidas, Levi Strauss & Co., and Burberry, utilize AI-powered analysis to manage inventory, optimize the supply chain, and adjust global distribution effectively (World Fashion Exchange, 2024). H&M has developed an AI-driven supply chain management platform in cooperation with Google Cloud. In this centralized platform, AI-based and machine learning algorithms are used to analyze the data coming from various sources like suppliers, warehouses and sales channels and to recommend adjustments in production or inventory allocation. Moreover, an advanced AI-based logistic system is beneficial for optimizing shipping routes and minimizing delivery times (DigitalDefynd, 2024a).

Trend forecasting, including demand prediction and trend analysis, is also a critical mechanism in fashion retailing. The use of modern technology and data-driven tools in trend forecasting eliminates the problems because of the unpredictability of consumer preferences and helps to envision emerging fashion trends. In this respect, Zara has implemented AI-powered analytics and predictive modeling to work on numerous data, including past sales, current purchasing patterns, weather forecasts, and social media trends, in order to make a demand prediction. With the use of this tool, it becomes possible to realize an increasing demand for a specific category in real time and include the change in the predictive modelling. Moreover, Zara uses AI tools to discover potential trends in order to incorporate them into its designs. For this aim, Zara monitors the social media platforms, blogs, and online forums, and it even employs sentiment analyses to understand the consumers' reactions towards the new trends (DigitalDefynd, 2024b). Similarly, H&M benefits from AI-driven technologies, particularly machine learning algorithms and natural language processing, to analyze the data from social media platforms and detect trending colors, styles, and materials early on (DigitalDefynd, 2024a). A case in point, the data scientists working at Levi Strauss & Co. discovered that baggy and loose silhouettes weren't just for the TikTok generation but for everyone after gathering data points from purchases made in 110 countries and 50,000 points of distribution, web browsing, retail partner sales, and its loyalty program (Bousquette, 2025). Trend forecasting based on modern technology has significant benefits for retail businesses. Indeed, Kering has reported that the accuracy of demand forecasting improved by 20% with AI-based demand planning (Balchandani et al., 2024).

The forecasted trends are taken one step further with the implementation of AI-based design processes. AI-powered curation across content and search was implied to be the main focus in the fashion industry, according to the McKinsey report (Balchandani et al., 2024). Specifically, generative AI has great potential in the design and creation processes since it can create images, visuals, and videos from scratch or from the current images. Moreover, it can draw sketches based on the trendy fashion attributes defined and the images provided. The popularity of AI-based design is increasing enormously, since generative AI is expected to add $150 billion to $275 billion in profits to the fashion industry by 2030 (World Fashion Exchange, 2024). Therefore, generative AI is widely used in both the design process and the establishment of fashion design campaigns. For instance, G Star Raw presented its first denim couture piece created by using a generative platform, Midjourney, which is produced in an in-house atelier in Amsterdam (Ryder, 2024). Valentino has employed a generative text-to-image AI tool to design all images in the "Essential" campaign in 2023 (World Fashion Exchange, 2024). As a part of its curatorial Genius division, Moncler collaborated with Maison Meta to release its first AI-powered campaign involving machine-generated images at the same time as Moncler's London Fashion Week spectacle (Ryder, 2024). Mango used AI-generated glam bots in advertising campaigns, reporting that the cost of that type of campaign is very low compared to working with a real-life model (Kato, 2024).

Modern technologies take a role in not only the creation and production activities but also in maintaining their values. In this regard, modern technology, specifically computer vision and machine learning, is used in the verification of the authenticity of products. For instance, Burberry benefits from image recognition and machine learning-based technology provided by Entrupy, having 98% accuracy in dealing with counterfeits (Marr, 2017). Similarly, the second-hand luxury retailer The RealReal makes product authentication by using machine learning and AI (Williams, 2024).

Finally, refining sustainable practices and building sustainable supply chain management is facilitated with the use of data-driven technologies. In this respect, AI, Big Data, IoT, and blockchain technologies can improve sustainability in many different ways, including facilitating recycling and reuse, optimizing management, and ensuring transparency and traceability in the supply chain. H&M benefits from an automated garment sorting system based on visual technology to automatically classify garment type, garment composition, and garment structure with an accuracy of 95% (H&M Foundation, 2024). Stella McCartney has teamed up with Google Cloud on machine learning and data analytics to achieve its sustainability goals (Donnelly, 2019). Levi Strauss & Co. has integrated AI into its supply chain to improve transparency and reduce the environmental impact of denim production (Zaytsev, 2024).

3.2 *Use of Modern Tools as a Part of Service Quality*

The retail industry provides not only products but also services, whose quality is important for consumers. Service quality increases customer satisfaction, sales, and profits. It was reported that 75% of shoppers were likely to spend more after receiving high-quality service (Balchandani et al., 2024). The changes made in services to increase shopping efficiency and flexibility for consumers have the potential to improve service quality. The interactive retail service through the use of modern tools, including RFID tags and smart touchpoints such as smart kiosks, smart price tags, smart carts, and unmanned shops, makes it possible for consumers to reach the required data and products easily and to increase shopping efficiency (Perumal et al., 2022). Self-service increases the pace of shopping, and the novel payment systems eliminate the need to carry cash or cards, as well as the check-out queues, by encouraging consumers to buy more (Pfeifer et al., 2023). Besides, especially the advanced options for product search, the self-service and self-checkout options reduce the labor cost for retailers. On the other hand, smart physical stores equipped with multiple cameras and shelves with weight sensors might track all the products and monitor the consumers (Chang et al., 2023). A touch screen in the self-service kiosk for the return items can integrate the retail system and transmit information acquired from consumers with only a few taps on the digital screen (Kim et al., 2020). Thus, the digitalization of some services also returns some additional valuable data for retailers, which results in making both real-time and future advancements. Modern tools based on AI, Big Data, IoT, RFID, and blockchain technologies can be used to improve service quality in-store and online activities, including store monitoring, self-service options, and product search and payment options.

In regard to store monitoring, AI-based computer vision provides data for retailers to develop informed visual merchandising strategies. Computer vision monitors consumers entering, exiting, and moving through different store areas (Soni, 2021). The customer movement might be captured with the creation of heatmaps that express the movement with different colors and hues (warmer colors indicating higher foot traffic, highlighted colors indicating the areas where consumers linger). Based on the data from computer vision technologies, the retailers might arrange product placement, store layout, staffing, and promotional strategies. The store arrangement can be optimized by analyzing the traffic to increase sales and customer retention rates. Yet, facial expressions, body language, and even gaze direction can be analyzed to understand consumers' reactions better and build informed decisions (Soni, 2021). Many retailers use this type of technology in their physical stores. For instance, M&S has collaborated with SymphonyAI Retail CPG to use computer vision to optimize store operations. The advanced AI software has been utilized for capturing real-time shelf images, rapidly identifying misplaced items or incorrect product assortments, and providing actionable insight to personnel (DigitalDefynd, 2025).

RFID tags are preferred by apparel retailers for store monitoring because of their additional potential in inventory management. RFID tags enable retailers to follow the items in stores to make further analyses about the hot spots or popular sections in stores to optimize the store layouts and product placements. Moreover, they interoperate with the inventory management and recommendation systems to guide consumers to their desired products. Zara, known for its superiority in inventory management, has utilized RFID technology for years in both the supply chain and in stores to ensure that popular items remain in stock. The department store Macy's has enhanced inventory accuracy and its potential to make personalized offers with RFID-tagged items (Invento RFID LLC., 2024).

Being the transmitters that can be located in stores, beacons are also used to increase service quality via store monitoring. The retailers utilize beacons to obtain data about visitors' flow, traffic, and hot spots in stores and to communicate with consumers via wireless signals like Bluetooth or Wi-Fi by sending location-based notifications. In this respect, Macy's has utilized beacons installed in its stores to maximize the traffic across its stores on Black Friday with the help of the application downloaded by consumers that sent push notifications to consumers when they approached the stores. Being a relevant example of out of fashion, Levi Strauss & Co. makes it easier to find their seats and reach food and drinks without any queues in Levi's Stadium (SDT Inc., 2022).

In regard to self-service, smart kiosks, self-checkout systems, and scan-and-go applications are common tools. For instance, the fashion retailers Zara and Uniqlo have installed sleek stations that work on RFID tags. The self-check-out system of Zara allows consumers to place the items in a bin to be priced without a need for a barcode scan. Using self-check-out systems Uniqlo reported that they have been used in 70–90% of transactions across markets and decreased the transaction times in half (Balchandani et al., 2024). M&S has offered a scan-and-shop application that allows consumers to scan the barcode of the items in the store and bag the items and then make the payment via the application by using a card, Apple Pay or Google Pay (M&S Sparks, n.d.).

Computer vision and image recognition technologies are employed for product search, which allows consumers to search for a product by using a photo. The photo is analyzed in terms of image, style, color, pattern and shape and matched with the products in the inventory. Even though the product search option can be used in both physical and online stores, it is favored by online retailers. ASOS has an AI-powered visual search shown as a camera icon in the search bar in its application, which has the capability of working with the photos or images uploaded by consumers (Arthur, 2017). Amazon allows consumers to upload an image of a product to find similar products, whereas Target's mobile application matches similar products by using both visual and voice searches (World Fashion Exchange, 2024).

Cashless payment, digital payment solutions, and self-checkout gained popularity with the adoption of e-commerce and mobile commerce. Near Field Communication (NFC) payment allows one to make payments with a credit/debit card and mobile wallets such as Apple Pay and Google Pay. QR-code-based payment is a safe and secure end-to-end digital payment technology (Gurusamy, 2024). Payment by link is another digital payment method that eases transactions. The digital systems have been enhanced with chatbot payments and biometric payments such as face recognition, fingerprints, and retina scans for authentication (Chang et al., 2021). In this regard, Prada and Hugo Boss offer payments by link through WhatsApp, email, phone, and other channels due to Adyen's Pay By Link feature (Samuely, 2024). Amazon stores use the AI-based voice assistant "Alexa" for Amazon stores. "Smile to Pay" is a biometric payment system introduced in 2017 by Alipay (Zhong & Moon, 2022).

Finally, cryptocurrencies such as Bitcoin and Ethereum, built on blockchain technology, have been adopted as a payment method to prevent double-spending while offering untraceability and anonymity (Xu et al., 2019). In this respect, Gucci became the first luxury brand to accept cryptocurrencies in 2021. Nowadays 70% of Gucci stores accept cryptocurrency (Loeb, 2022). London-based fashion store Philipp Plein has opened a crypto-concept store on London's Old Bond Street, where consumers can purchase unique fashion NFTs by using cryptocurrencies (Halliday, 2022). Similarly, French luxury fashion brand Balenciaga announced that it would accept cryptocurrencies Bitcoin and Ethereum in some brick-and-mortar stores in the USA (Business Today Desk, 2022). Ralph Lauren has begun to accept cryptocurrency in a partnership with BitPay (McDowell & Schulz, 2023).

3.3 Use of Modern Tools as a Part of Mass Customization

Referring to providing customized goods or services that best meet the customers' needs and requirements, mass customization is provided with the presentation of tailored products and services and the implementation of cocreation activities. Mass customization increases the value of products and the quality of services (Ng et al., 2021). Consumers receiving the most convenient products in a shorter time period become satisfied with the shopping experience, and they don't need to return the products because of fit issues (Tawira & Ivanov, 2023). Additionally, gaining the advantages of specialized offers and promotions makes consumers feel special and more satisfied, which can lead them to be loyal consumers in the long run. Higher customer satisfaction and a reduced number of returns increase sales and profits on the one hand and improve retailers' commitment to sustainability on the other hand. For these reasons, "Personalized design and production" is shown as one of the top three objectives considered by fashion retailers in a report by McKinsey. In the same report, tailored product recommendations and personalized promotions offered by Kering Group are recorded to increase the order value by around 15 to 20% (Balchandani et al., 2024). The integration of AI, Big Data, IoT, VR, AR, and the relevant technologies into modern tools, including smart assistants and smart mirrors, increases the mass customization capabilities in the recommendation of the most convenient product, promotion, or service for consumers. Besides, these technologies allow consumers to take part in cocreation activities.

Smart assistants such as chatbots and voicebots are text-based or voice-based recommendation tools whose capabilities and intelligence are enhanced with the integration of AI-based technologies such as natural language processing and speech recognition (Jiang et al., 2022; Rohit et al., 2024). The chatbots or voicebots might respond to consumers in real time by providing online support services, which enable consumers to make product searches and online purchases and track the status of orders (Rohit et al., 2024). Smart assistants have been widely used by fashion retailers for years. As one of the earliest chatbots in the fashion industry, the Facebook Messenger chatbot by Tommy Hilfiger, TMY.GRL, was created in partnership with the artificial intelligence platform msg.ai. and used to offer style advice and product recommendations depending on information about users' personal style preferences (Arthur, 2016). Using IBM Watson's machine learning technology in the background, The North Face's virtual assistant provides personalized recommendations based on the answers from consumers about where the product will be used (Sen, 2018). "Nike Maker Experience" offers consumers the ability to request shoes in different sizes and colors with voice commands (Mathews, 2024). Even there are some companies in fashion retailing that have built their business models on providing style assistance to consumers. In this respect, Stitch Fix is a personal styling service that utilizes an NLP-based recommendation system. The advanced AI algorithms used by Stitch Fix analyze consumers by using data about style preferences, past purchases, and current fashion trends (Vogue Business in partnership with Stitch Fix, 2024).

The capabilities of smart assistants have steadily increased to improve customer satisfaction and enhance the consumer experience. In this respect, some features, such as social media interaction, have been added to these services. For instance, H&M has collaborated with the messaging platform Kik and built a chatbot on this platform where the chatbots could interact with each other and the users might send invitations to friends during their chatbot conversations (Samuely, n.d.).

Smart mirrors and smart dressing rooms are another type of mass customization tool, which presents a more interactive service to consumers. These tools are superior to chatbots and voice bots in fashion product recommendations, as they provide not only textual information but also visual information in terms of fit, size, and looks. They allow virtual try-ons and provide product recommendations based on a more personalized fit, ultimately leading to greater success. Moreover, they reduce the time requirement for try-ons, product search, and payment resulting, in service quality and better customer experience (Alanazi & Alenazi, 2023).

Smart fashion mirrors, also called magic mirrors, digital mirrors, or interactive mirrors, utilize a two-way mirror with cameras, electronic displays, and sensors behind the glass with augmented gesture recognition technology (Ogunjimi et al., 2021). As mentioned previously in Sect. 2.2, they are using different technologies, but the ones based on AR are the most common types for virtual try-ons, which makes it possible to try fashion products without wearing them.

Smart mirrors and dressing rooms gained huge popularity among fashion brands. ASOS has allowed customers to check the fit of clothes on various body types via the "See My Fit" service during the coronavirus period (Rhee & Lee, 2021). Timberland has used an AR-based tool that allows consumers to browse the catalog, choose their sizes, and superimpose Timberland clothing and accessories on them (de Amorim et al., 2022). A fashion retailer, Goddiva, has collaborated with Zyler, a company specializing in AI-powered virtual try-on technologies, and enabled online shoppers to try on garments after they upload their body measurements and/or a photograph (Ward, 2024).

The smart mirrors and dressing rooms are integrated with special features. For instance, Topshop introduced a virtual fitting room using Kinect, a motion-sensing technology (Mattan, 2024). The European fashion retailer Zalando added the heat maps feature to its smart dressing application, in which consumers create their 3D avatars by entering their weight, height, and gender. The fit of different sizes is visualized by using the heat map function, indicating the narrower or wider areas (Ronnie, 2023). Integrating computer vision, machine learning, AR, and recommendation models at the same time, Nike Fit offers exact shoe sizes for consumers based on scans of clients' feet with a smartphone (Mathews, 2024).

The features of smart mirrors are enhanced with additional technologies such as RFID, IoT, Big Data, and AI technology that make it possible for smart mirrors to reach inventory data, show various sizes and colors, and offer alternative styles and outfits or personalized recommendations to encourage consumers to make additional purchases (Alanazi & Alenazi, 2023; Lee & Xu, 2020). Rebecca Minkoff uses smart fitting rooms equipped with RFID, which allows consumers to interact with touchscreens to request different sizes, colors, or styles. (Invento RFID LLC., 2024). Zara utilizes a similar type of smart mirror equipped with RFID systems, with an additional feature of giving outfit suggestions according to the product selected. The Zeekit VTO lets consumers mix different combinations of tops and bottoms on their avatar and create dozens of clothing ensembles in minutes to be purchased or saved in a wish list (Tawira & Ivanov, 2023).

Smart mirrors and dressing rooms can be transformed into socially interactive dressing rooms. The socially interactive dressing rooms basically have access to in-store services and/or social networking websites. In the interactive dressing rooms equipped with RFID music tags, it is possible to play music that matches the garment style or play the songs of celebrities associated with the product or brand. For example, a customer trying on jeans advertised by a pop star could trigger the pop star's music to play in the dressing room. Besides, consumers might access contacts in popular video conferencing or instant chat applications by social networking websites (e.g., Skype, Yahoo Messenger, and Facebook) in socially interactive dressing rooms (Kim et al., 2017). Consumers can talk to their friends to get feedback on their clothing options or share their images or videos on social media (Lee & Xu, 2020).

The COS stores of H&M in the USA offer virtual try-ons and personalized styling recommendations by using smart mirrors, which operate as a digital display in connection with smartphones via Bluetooth or Wi-Fi. This mirror is capable of allowing consumers to take selfies that are virtually integrated into the H&M catalogue (SDT Inc., 2022). The AR application Kohl's Virtual Closet enabled users to virtually try on activewear and athleisure wear items and capture photos by using Snapchat's new Selfie Lens feature. Levi Strauss & Co. introduced an online co-watching video application utilizing AR and the digital tool "Squad," having the feature of allowing users to shop together with their friends (Kang et al., 2023).

On the other hand, the try-ons are not specific to physical products only. They can also be used with virtual designs that do not exist physically. For instance, Bershka has collaborated with FFFACE.ME to introduce the first semi-digital wearables collection, Wearable Arts, enabling the physical clothing to be augmented with digital elements by using Instagram filters (Taggart, 2024).

As mentioned earlier, mass customization is also possible through the cocreation of products with consumers. In fact, many customers do not like standardized products (Pillai et al., 2020), and they feel appreciated when they assume the role of creator. Involving consumers in designing new products increases their acceptance (Kim & Choo, 2023). For this reason, cocreation suggests retailers strengthen their role in sustainability as they provide a lower carbon footprint because of accepted products and reduced clothing returns. (Tawira & Ivanov, 2023; Kim, 2021). Even though cocreation is not so widespread yet, some fashion brands present cocreation in specific product categories. For instance, Left Shoe Company provides consumers with made-to-measure shoes by allowing consumers to select the shoe style and material and have their feet scanned at one of their pop-up stores (Dacko, 2017). Adidas customizes its shoes according to consumers' foot shape, size, and cushioning needs by utilizing 3D printing and generative AI technology (Mearian, 2015). Besides, AI-powered tool Zegna X Creator by Zegna allows consumers to create their own designs and reach the available fabric options for these designs via its direct connection to updated fabric catalogs (Salamone, 2023).

3.4 *Use of Modern Tools as a Part of Consumer Experience*

Defined as "*the total of feeling, perception, and attitudes formed during the entire process of decision making through interaction with people, objects, processes, and environment*" (Lee & Xu, 2020), consumer experience occurs at any type of physical and digital platform and every stage of purchasing, including pre-purchase, purchase, and post-purchase steps. Presenting a positive experience to consumers is advantageous for retailers because it leads to customer engagement, and engaged consumers make frequent purchases, more transactions per purchase, and become more passionate about the brand or retailer. Engaging consumers helps to build long-term relationships and make firms achieve a sustainable competitive advantage (Sung, 2020). The retailers improve consumer experience by increasing the utilitarian and hedonic benefits for consumers. Additionally, since consumer experience is conceptualized to cover both entertainment and adventure (Mohd-Ramly & Omar, 2017), retailers put efforts into providing innovative selling initiatives and establishing novel and enjoyable marketing activities.

Modern technologies such as AI, IoT, blockchain, robotics, and particularly the digital technologies AR and VR have a huge potential to create an entertaining experience. Moreover, the rapid growth of VR and AR technologies, which resulted in the rise of Web 3.0 and the creation of a 3D digital universe, Metaverse, can be used for this purpose (Muhammad Sohail Jafar et al., 2024). Having the meaning of beyond universe, the metaverse has integrated physical and virtual reality. Since it is a computer-generated environment where multiple users from different locations can interact with each other simultaneously, the metaverse makes it possible for users to immerse themselves in shared and real-time experiences (Kim et al., 2023).

Specifically, VR has been employed for preparing a joyful and unique experience with the establishment of virtual stores and storefronts, special activities, and virtual fashion shows. An Australian fashion retailer, Myer, launched a VR departmental store to offer an immersive shopping experience and enable consumers to browse and purchase items online (Jin et al., 2021). Rebecca Minkoff offered a virtual store where consumers can enter the VR world and shop as if they are in a real shopping setting. Consumers also have access to in-depth videos in this virtual store (Kim et al., 2021). Tommy Hilfiger and Allison Mitchell have created a 360-degree virtual storefront in collaboration with Obsess, a VR platform enabler (Jin et al., 2021). Adidas launched Adidas TERREX products and introduced them with a VR experience for its customers involving following climbers through their ascent (Farah et al., 2019). TopShop has offered a VR fashion show with a 360-degree view of the products (Jin et al., 2021).

Because of its potential to change society, commerce, entertainment, education, and social interaction (Kalyvaki et al., 2023), fashion brands showed great interest in the metaverse. In this respect, they either opened their own stores or collaborated with metaverse platforms to promote their brands and enhance consumer engagement, satisfaction, and overall experience. Gucci has launched its metaverse platform in Roblox to mark its 100th anniversary and created a virtual garden exhibition where users can explore and buy digital collectibles (Afolabi, 2023). On this platform, the virtual version of Gucci's Dionysus bag was sold for $4115, higher than its real-world version (Malarvannan, 2023). Moreover, Nike launched its Web 3.0 marketplace platform, SWOOSH, in 2022, which allowed users to purchase virtual products (Fakhry & Nasr, 2023). Burberry launched its virtual Lola Bag Handbag collection with unique designs on the Roblox platform (Afolabi, 2023). The fast fashion retailer Zara entered into the digital fashion market through a collaboration with Zepeto, a South Korean metaverse platform with nearly 300 million users. Zara launched a virtual clothing collection as part of its Lime Glam campaign. This collection also included physical pieces that can be worn in real life. Zara also introduced its new channel on Snapchat, where special AR-based filters were created as a part of the Life Glam campaign (Malarvannan, 2023). Forever 21 created the virtual world "Forever 21 Shop City" in the metaverse. In this virtual world, not only were the products sold, but also users were allowed to create and manage personalized stores and to hire non-player character employees (Takahashi, 2021).

Other brands presented different experiences in addition to offerings related to fashion. For instance, the American fashion brand Ralph Lauren established the themed event Ralph Lauren Winter Escape, which enabled the users to enjoy winter activities, including ice skating and drinking hot chocolate, and browse some vintage fashion from the Ralph Lauren Digital Collection (Adegeest, 2023). Gucci launched the Gucci Vault on the Sandbox platform, which is the virtual creation of the Gucci store in Milan. The users can purchase NFTs and vintage fashion pieces, learn more about the history and culture of Gucci, and interact with other Gucci enthusiasts by playing games (Afolabi, 2023). To attract the attention of younger users by inviting them to cocreation activities, Tommy Hilfiger presented the "*Tommy x Roblox Creators*" collection, through a user-generated content partnership. The collection involved 30 digital fashion items designed by users involving the brand's style and their creative output (Lockwood, 2021).

The other brands focused more on entertainment. Nike launched Nikeland on the Roblox platform, where users can engage in sports games, enter competitions, and interact with other Nike enthusiasts (Afolabi, 2023). Some brands partnered with gaming companies. Well-known French luxury fashion house Balenciaga partnered with Epic Games for the game Fortnite in the metaverse. Within this partnership, some Fortnite characters were adorned with Balenciaga's iconic products, and the users were presented with some virtual Balenciaga products within the game, Fortnite (Kelly, 2021). To celebrate the brand's 200th anniversary, Louis Vuitton joined Web 3.0 and introduced a mobile game called "Louis the Game" to be downloaded on IOS and Android in collaboration with digital artist Beeple. In the game, the avatars are customized with LV prints (Malarvannan, 2023). Burberry released a limited-edition NFT collection for the "Blankos Block Party" multiplayer game in collaboration with Mythical Games (Afolabi, 2023).

In parallel with the developments in Web 3.0, the metaverse, and blockchain, NFTs have also gained significant popularity. Even though NFTs are digital content themselves and they serve to dematerialize physical assets, they can also be used as a bridge for materializing virtual ones (Park et al., 2023), with the rise of the phygital products, which correspond to the NFTs and their physical duplicates.

As in the case of phygital products, a hybrid consumption experience, "phygitality," became possible with the integration of physical, human, digital, and media content elements; platforms; technologies; and extended realities together in combination with the digital and physical experiences (Alexander & Varley, 2025).

In this respect, Puma entered digital fashion marketing with the platform "Black Station," where consumers can purchase NFTs and convert them into physical goods (PUMA, n.d.). Gucci partnered with the company Superplastic, producing animated celebrities. Within this collaboration, 10 unique NFTs having physical ceramic sculptures have been developed and sold via Gucci Vault (Schulz, 2022). Dolce & Gabbana released 9 NFTs with their physical versions under the name "Collezione Genesi NFT collection" in 2021 (Thomas, 2021). Prada introduced the initiative "Prada Timecapsule," dropped many NFTs on a monthly basis, and released the phygital tank tops as the 42nd NFT collection in June 2023 (Sarvaiya, 2025).

Moreover, fashion brands collaborated in some campaigns for NFTs. Adidas and Prada collaborated for a limited-edition NFT collection called Adidas for Prada Re-Source (Malarvannan, 2023). Adidas also collaborated with Italian luxury brand Moncler in Web 3.0 and launched "The Art of Explorers," bringing the characteristic design details of both brands (Moncler, 2023). Before acquiring the digital fashion company RTFKT, which creates digital sneakers, Nike launched the Nike Air Force 1 forging event in 2023 in collaboration with RTFKT. In this collaboration,

Nike presented an NFT collection including 10 items, two of which were designed by Takashi Murakami. What made this event special was that anyone who bought an NFT during the event could also get a physical copy of the NFT (Afolabi, 2023).

Digital fashion created new brands, such as the Fabricant and RTFKT. Launched in 2018, the Fabricant allowed users to cocreate digital apparel (Johnson, 2024). Similarly, RTFKT, which was founded in 2020, collaborated with iconic artists and designers and became very popular with its CloneX NFT collection and was then purchased by Nike (Johnson, 2024).

Metaverse also presented many platforms for fashion exhibitions. In this respect, the first Metaverse Fashion Week was established on 24–27 March 2022 on Decentraland. Many fashion brands, such as Dolce & Gabbana, Etro, and Tommy Hilfiger, have established digital shows and exhibitions at this fashion event. Besides many activities and events were organized, such as musical performances, gatherings, openings, and after-parties. Plenty of opportunities were presented to shop for NFTs and wearable outfits (Pasquinelli, n.d.). In this respect, 108,000 guests attended the Fashion Week and bought over 7000 digital garments. Specifically, Dolce & Gabbana has established a catwalk and presented 20 designs on cat-faced avatars. Tommy Hilfiger presented NFTs of a virtual version of its Spring 2022 collection (Zara & Hobbs, n.d.). Selfridges opened its flagship store inspired by the retailer's Birmingham store and offered visitors a tour of past works and highlights. Metaverse Fashion Week 2023, with the theme "Future Heritage," was established on 28–31 March on Decentraland, supported by the Ethereum blockchain. There were many exhibitors at the event, including the brands Coach, Tommy Hilfiger, and Diesel, alongside web natives DressX, The Fabricant, and the Institute of Digital Fashion, and more than 60 designers and artists. Specifically, DKNY offered an immersive experience, with a four-floor building consisting of a boutique, an art gallery, and a rooftop lounge with a view of the city skyline. Tommy Hilfiger arranged a competition among the designers, Adidas established a party and a fashion show, and Diesel hosted a DJ set and a party where participants could watch a digital fashion show on display (Pasquinelli, n.d.). It is planned to have the third fashion week, MVFW, on 9–12 April 2025 with the participation of brands and more than 47 global designers with the theme "Infinite Identities." In the event, the designers are expected to present innovative animated collections on the runways, and 16 interactive design showrooms are expected to be established in two thematic zones, "Augmented Self" and "Storied Self," highlighting the phygital fashion and global narratives with identity-driven collections (Lezama, 2025).

4 Modern Tools within Consumer Perspective

In this section, the consumers' approaches towards smart retailing and individual modern tools are investigated with the findings of the most recent empirical studies published between the years 2020 and 2025. The studies summarized in this section mostly employ the adoption and technology theories, including the Technology

Acceptance Model (TAM), the Unified Theory of Acceptance and Use of Technology (UTAUT) models, the Stimulus-Organism-Behavior (SOR) model, the Theory of Consumption Value (TCV), the Elaboration Likelihood Model (ELM), the Expectation Confirmation Theory (ECT), the Affordances Actualization Theory(AAT), the Social Capital Theory(SCT), the Media Richness Theory (MRT), the Uses and Gratification Theory(UGT), the Attribute- Evaluation-Outcome model (AEO), and the Information Sharing Theory (IST) but the studies changed in terms of scope, the conceptual models applied, and the sample characteristics. Whereas the first subsection is dedicated to smart retailing in general, including the application of modern tools in all types of industries other than fashion and apparel, the second subsection includes only the studies established for the apparel and fashion industry in order to gain specific insight about the similarities and differences with the other industries. The section ends with the third subsection, which concentrates on the impact of consumers' characteristics on behaviors.

4.1 *Consumer Approaches Towards Modern Tools in Smart Retailing*

This section is divided into two subsections: the first one focuses on investigating the motivations behind consumer behavior, while the second one compares different tools and reveals consumer expectations regarding various types of tools. The section concludes with a recommendation based on the findings presented in these two sub-sections.

4.1.1 Motivations Towards Modern Tools in Smart Retailing

Numerous studies focused on the identification of motivations beyond the acceptance, adoption, and usage intention of smart retailing and modern tools. Usefulness, ease of use, and perceived enjoyment were proposed as the most important motivations leading to the acceptance and adoption of smart retail technology (Perumal et al., 2022; Chang & Chen, 2021). Silva et al. (2023) confirmed the positive impacts of the perceived usefulness and perceived ease of use on attitudes toward using chatbots in Portuguese. Gümüş and Çark (2021) revealed that the perceived ease of use and usefulness were influential on the behavioral intention, adding that the perceived enjoyment affected the customer experience related to the chatbots in Turkey. Similarly, Rehman et al. (2023) confirmed that enjoyment could be considered as a part of emotional experiences, which had a positive impact on consumers' purchase intentions towards a social robot in retail settings. Concentrating on the AR applications in shopping malls, Saprikis et al. (2020a) confirmed the positive impacts of performance expectancy followed by enjoyment on behavioral intention in the mentioned order. Jiang et al. (2023) revealed that perceived usefulness had the highest impact on the attitude, whereas perceived playfulness had a significant impact on continuance intention to use virtual try-on applications for shoes in China.

The researchers, particularly focusing on user experience, have examined smart retailing and modern tools from both utilitarian and hedonic perspectives. Lee et al. (2021) examined the social diffusion of artificial intelligence-based voice assistant systems and showed that hedonic motivation and compatibility were influential on satisfaction, which led to the use of continuance and intention to recommend. Mishra et al. (2022) demonstrated the effect of utilitarian and hedonic attitudes of smart voice assistants on the use and recommendation intention. Chattaraman et al. (2024) revealed that AI-based mobile decision bots enhanced the consumer journey by satisfying both the hedonic and utilitarian expectations. Among researchers focusing on hedonic and utilitarian aspects, modern tools and smart retail strongly addressed the significance of hedonic motivations. Chang et al. (2023) showed that hedonic motivations had a higher impact than utilitarian motivations for purchase intentions in smart stores. Similarly, Ferber and Vaziri (2024) implied that the hedonic qualities contributed positively to customers' intention to use a robot. Finally, Kim and Kim (2024) revealed that the extrinsic value and epistemic value of meta-commerce consumption caused higher satisfaction, but it was the hedonic value that resulted in continuous usage intention.

Along with these primary motivations, it has been demonstrated that social influences were also influential on consumer behavior. Lin (2022) employed the UTAUT model and showed that consumer attitudes towards intelligent unmanned convenience stores in Taiwan were positively influenced by social influence. Sharma et al. (2022) confirmed the positive impact of social influence on the adoption of autonomous decision-making processes in smart stores and added that collectivism strengthened the relationship between social influence and customer attitude. Focusing on QR-based payment, many authors revealed that social influence was one of the primary motivations for the adoption and use intention of these tools (Baskoro & Amini, 2020; Le, 2022; Widayat & Masudin, 2023; Tu et al., 2022; Chang et al., 2021). Pillai et al. (2024a) indicated the significance of social influence on the shopping intention in metaverse platforms. Patil and Pramod (2022) showed that the metaverse blended retail purchase intention was influenced more by social gratification than hedonic gratification since the metaverse retail could bring networking and a sense of belonging to a community to retain loyal consumers.

The innovative features of the modern tools were emphasized in consumers' adaptation and experiences. Perumal et al. (2023) indicated that the perceived novelty had the highest positive impact on shoppers' attitudes towards smart retailing technologies in Malaysia. Adapa et al. (2020) implied that the perceived novelty and the perceived advantage determined the perceived shopping value of smart retailing carts. In another study established on smart retailing carts in North America, the perceived novelty and the perceived compatibility were determined as two main factors enabling the consumer's intention to use these tools (Fazal-e-Hasan et al., 2021). On the other hand, some researchers examined the impact of novelty through the lenses of consumers and thus investigated the effect of consumer innovativeness. In this respect, Oyman et al. (2022) showed that consumer novelty seeking had a positive and direct effect on the positive perception of the use of AR in mobile applications for cosmetic products. Lee et al. (2021) confirmed that personal innovativeness was linked to recommending and reusing the chatbot. Finally, Pillai et al. (2020) found out that innovativeness had a positive impact on both the perceived ease of use and perceived usefulness of AI-powered automated retail stores in India, together with optimism.

Personalization and customization were also found to be influential on consumer behavior towards smart retailing and modern tools in a positive manner. Extending the technology readiness and acceptance model with AI context-specific constructs, Ameen et al. (2022a) showed customization and interactivity as the significant predictors of shopping intention in AI-powered automated stores and stated that the personalization improved loyalty in India. The perceived personalization and responsiveness were confirmed to play an important role in the perceived playfulness of the mobile AR-based tool for shoes in another study (Jiang et al., 2021). Concentrating on the smart cocreation experience, Roy et al. (2022) indicated that perceived personalization had the strongest effect on smart serviscape, whereas Shen et al. (2025) revealed that perceived personalization was among the most important components of service quality.

However, the inclusion of trust and risk as motivational components in the studies suggests that modern tools and technologies did not instill confidence in consumers. Fazal-e-Hasan et al. (2021) found that the perceived risk had a negative impact on the intention to use smart retail. Pillai et al. (2020) showed that insecurity negatively impacted the perceived usefulness of AI-powered automated retail stores in India. Lin (2022) added that patronage intentions towards unmanned stores were negatively affected by perceived risk. Based on the interviews with eight participants, Dev and Dev (2023, April) revealed that perceived privacy risks could decrease the adoption rate of chatbots. Silva et al. (2023) confirmed that trust was influential on perceived usefulness, user satisfaction, and attitudes toward using chatbots. Ngo and Nguyen (2021) showed that perceived security was one of the most important factors that has driven the use of QR payment in Vietnam. Trust was considered an important factor in novel payment systems in many studies (Türker et al., 2022; Rafferty & Fajar, 2022;Kilani et al., 2023).

On the other hand, it has been observed that consumers' feelings of inadequacy lead to avoidance of modern tools. In this respect, Alam et al. (2024) showed that the user intentions for service robots were influenced by technological self-efficacy, together with habitual behavior and subjective norms, and Zhong and Hamouda (2024) revealed that the intention to participate in the Metaverse was influenced by self-efficacy.

Some researchers associated the factor of self-efficacy with feelings of anxiety or the term perceived complexity. Lee et al. (2021) indicated that the technology anxiety indirectly caused a lower intention to recommend the AI-based voice assistant systems. Moreover, Adapa et al. (2020) stated that the perceived value was influenced negatively by the perceived complexity of smart retail tools. Having examined shoppers' response to beacon-triggered promotions by employing tests in Serbia, Croatia, and Bulgaria, Đurđević et al. (2022) revealed that the ease of participation was positively influential on the acceptance and redemption of beacon technology.

4.1.2 Expectations from Modern Tools in Smart Retailing

The expectations of consumers varied depending on the tool. Main expectations from chatbots, especially the text-based and voice-based decision bots, were the compatibility of data derived by these decision bots, the anthropomorphism, and the social presence. In a study established by Araújo and Casais (2020), the adoption of chatbots as shopping assistants in e-commerce by Portuguese respondents between 18 and 54 years old was examined, and the compatibility of the recommendation was found to be the most important factor in the adoption. Aiolfi (2023) demonstrated that the task-technology fit of smart speakers enabled consumers to use these tools. In a similar vein, He et al. (2024) implied that user satisfaction increased with the diversity and accuracy of the recommendations. Comparing two types of decision bots by using the Elaboration Likelihood Model, Chattaraman et al. (2024) revealed that the decision bots providing attribute-level information were favored more than the alternative-based decision bots, especially by the consumers with low product knowledge. Flavián et al. (2023) compared the influence of traditional text-based recommendations with voice-based recommendations and showed that voice-based recommendations were effective in altering consumer behaviors and provided higher credibility and usefulness. In addition, Rohit et al. (2024) revealed that voice assistants were more effective than text-based chatbots in terms of both cognitive and affective engagement, especially for experiential products. It was also implied that consumers got engaged with the text-based chatbots for searching product displays, however.

Nevertheless, it is agreed upon that the consumers still preferred to engage with humans instead of chatbots because these tools were impersonal and lacked the human touch. In fact, Sidlauskiene et al. (2023) demonstrated that the personalization of chatbots would be positively perceived, and consumers would have a higher willingness to pay a higher product price when the chatbots had anthropomorphic characteristics. Kim and Hur (2024) further indicated that the personalization and anthropomorphism of AI-based chatbots significantly increased perceived competence and warmth, which might lead to empathy and consumers' willingness to use AI chatbots. Anthropomorphism was affirmed to increase perceived competence (Song et al., 2024).

Social presence and social interactions were found to be appreciated by consumers. As a matter of fact, Tsai et al. (2021) observed that high social presence enhanced with the anthropomorphic profile design led to higher consumer engagement in the end. Lee and Park (2022) revealed that a parasocial relationship with an AI-based shopping bot increased continuance usage intention through perceived communication quality in Korea. Similarly, Hsieh and Lee (2021) confirmed that the parasocial interactions together with media richness were the key determinants determining the continued usage and purchase intentions via AI assistant-enabled smart speakers. Akdim and Casaló (2023) studied the perceived value of AI-based recommendation services, particularly voice assistants, and found out that the social presence improved the benefits, personalization, compatibility, and convenience. The social presence and perceived expertise of AI chatbots were presented to motivate users to process information deeply (Pham Thi & Duong, 2025). On the other hand, Yang (2024) focused particularly on the effect of humor on satisfaction and demonstrated that the humor in chatbot interactions positively affected customer service satisfaction via their interestingness and competence. Nevertheless, the use of humor was recommended in only low-severity situations, but not in high-severity ones.

In addition to these, it was also interesting to see that the gender of the chatbots and recommendation systems was discussed in the literature. Ahn et al. (2022) investigated the effects of gender stereotypes on the evaluation of the AI-based recommendations and revealed that the male AI agents were more competent and the female AI agents were warmer. In connection with this, the male agents were more positively accepted when a recommendation was made for a utilitarian product. On the contrary, a female agent was preferred when a recommendation was needed for a hedonic product. Flavián et al. (2023) confirmed the effectiveness of the chatbots with male voices by using Media Richness Theory.

As in the case of chatbots, anthropomorphism, personalization, and social presence were the topics that were expected from the service robots. On the other hand, the consumers showed more skepticism towards service robots, probably because they were feeling more anxious about the complexity and autonomy of robots. In this respect, Schepers et al. (2022) found out that the emotional response of the users was more positive for the feeling-AI-based service robots than the thinking-AI-based and mechanical-AI-based robots. Albrecht and Honekamp (2022) examined the consumers' perception of two types of humanoid service robots: the one using deep acting (evoking thoughts, images, and memories) and the one using surface acting (faking the displayed emotions and shape) and stated that deep-acting robots were more successful than surface-acting robots. Employing the TAM model as a framework, Shehawy et al. (2025) showed that the attitudes towards robots were influenced by the appearance, capabilities, and personalization features of robots and interaction. In another study, the anthropomorphic assistant was stated to reduce the psychological distance (Jham et al., 2023) because the anthropomorphism, autonomy, and ability of service robots were positively associated with perceived usefulness (Li & Wang, 2022). On the contrary, Frank and Otterbring (2024) revealed that the high autonomy of AI-based tools had a negative effect on the adoption intentions because of creating a powerlessness for consumers.

Most of the studies established on AR and VR applications assessed the features of different applications. Álvarez Márquez and Ziegler (2020) evaluated an AR-based recommendation system that can be employed in physical stores. The system had the ability to provide critiques about the products in stores and offer alternative products available online. The application was found useful since it provided recommendations considering the visual similarity and technical attributes. Nonetheless, it was implied that the consumers usually made their final decision on the physical item and ignored the online recommendations. Zimmermann et al. (2023) compared the two mobile applications, the AR-based shopping assistant (ARR) and the AI-enhanced AR-based shopping assistant (XARS), in brick-and-mortar shopping scenarios and revealed that both ARSS and XARS were trustworthy. Moreover, both provided the necessary information, increased the shopping pace in an interactive manner, and enabled cross-selling/upselling at the point of sale. However, ARSS was sometimes low-functioning. In another comparison of AR-based applications and AI-based chatbots, Ameen et al. (2022a) showed that the augmentation property positively affected the body image and self-esteem of Generation Z female consumers. Pfeifer et al. (2023) compared ARSGs (augmented reality smart glasses) with AR on touchscreen devices and found that ARSGs were superior to AR on touchscreen devices in terms of immersion and mental intangibility. ARSG was stated to provide greater freedom of movement and a stronger sense of proximity to the products. In addition, De Amorim et al. (2022) assessed the AR-based reality shopping assistant developed by using HoloLens glasses and implied that media richness, including the amount and variety of information, and the ability to provide some immediate feedback to users were important factors for consumers.

Some studies compared the AR and VR applications. Kim et al. (2021) implied that VR provided a more engaging and enjoyable shopping experience than AR, but it was also more complicated and expensive. Xi et al. (2024) compared AR with VR in terms of perceived enjoyment, perceived ease of use, and perceived usefulness. The researchers revealed that AR was more useful but more difficult to use, whereas VR was more enjoyable. Concentrating on the 2D displays and 3D VR environments, Kang et al. (2020) revealed that informativeness and playfulness were important and 3D environments were appreciated more instead of still images and videos. Nunkoo et al. (2024) compared the metaverse (3D environment) with smartphone applications (2D environment) and implied that consumers preferred using smartphone applications over the metaverse because of habitual behaviors, convenience, and their easy-to-use interfaces. The authors added that 3D environments were more suitable for appealing products because of their potential to enhance immersiveness.

The expected features of VR and metaverse platforms were investigated in more detail in literature. In this respect, Balakrishnan et al. (2024) showed that the affective expectations of users were higher than the cognitive expectations, such as informativeness and design from the metaverse. Colorful designs and vividness enhance the positive experience (Balakrishnan et al., 2024). In another study, vividness was found to have a significant positive effect on perceived usefulness and perceived enjoyment, which had a crucial role in behavioral intentions (Kim et al., 2021). Warden et al. (2024) focused on the servicescape attributes in metaverse platforms and revealed that music, crowding, and noise were the most influential factors on emotions. Chen et al. (2024) implied that the visual, kinesthetic-based, and haptic-based stimulations enhanced multisensory experience and spatial presence and reduced product uncertainty and information overload in AR-based metaverse platforms. Moreover, Han et al. (2020) revealed that body ownership and control resulted in the adoption and utilization of VR technology. Besides, the sense of immersion and imagination was shown to enhance security and privacy (Pillai et al., 2024b). Finally, the perceived similarities and vividness followed by interactivity were revealed to serve as catalysts for the flow experience and telepresence feeling (Muhammad Sohail Jafar et al., 2024).

Since VR-based applications and the Metaverse enable more social interaction unlike all other applications, topics such as social experiences, symbolic value, and brand image, as well as the features and realism of avatars, were discussed in regard to expectations in the literature. Establishing a longitudinal analysis, Chakraborty et al. (2024) showed that the shared experiences significantly influenced the continued use of metaverse-based virtual stores. The social presence was shown to be an enabler, whereas realistic and symbolic threats related to concerns about the integrity of a value system were inhibitors of willingness to use the Metaverse (Kumar & Shankar, 2024). In parallel with this, Balakrishnan et al. (2024) implied that aspirational tendencies are the reason for using the metaverse. Concentrating on the NFT luxury brand marketing in the metaverse, Sung et al. (2023) drew attention to the impact of social value, authenticity, and scarcity of NFTs on purchase intention. The authors added that the closeness of avatars to users' ideal self-image resulted in positive attitudes. Kim et al. (2023) confirmed that the intention to use Metaverse was increased by avatar realism. The avatar realism was also stated to lead to a greater psychological distance because of the uncanny valley effect, however. Kalyvaki et al. (2023) worked with a sample of 214 active users participating in the Second Life to explore avatar attachment and showed that 75% of participants defined their avatar as an extension of themselves, 90% cared about the appearance of avatars, and 84% customized their avatars.

Based on the findings explained in Sect. 4.1.1 and 4.1.2 above, the authors provided some suggestions in order to eliminate the barriers and improve the acceptance, satisfaction, and experience.

As the difficulty in use was seen as a barrier in some studies, it was advised to simplify the modern tools. In this respect, it was recommended to craft user-friendly interfaces and integrate on-screen instructions in AR try-ons, and add tailored linguistic and cultural preferences in chatbots. Moreover, it was recommended to provide some form of education on how to use these tools. Incorporating trivia, quizzes, and enjoyable content was suggested as a way to teach users about modern tools.

The other barrier was security-related issues, and therefore it was advised to increase transparency provided and guarantee the protection of private data via the enforcement of robust privacy policies. Specifically, the chatbots were advised to be equipped with the incognito mode to handle requests involving private information more securely. Moreover, it was suggested to inform the users that the audio logs could easily be deleted from interaction history.

On the other hand, the functionalities of the modern tools were suggested to be improved by providing unbiased, high-quality information along with timeliness and synchronicity, delivering contextually accurate responses, and tailoring product recommendations. Besides, it was suggested to employ face recognition and image recognition systems to lower the shopping time. Customization was another suggestion to improve the functionality. Specifically, it was implied that the conversation with the chatbots should be enhanced by some personalized suggestions considering the purchase history and characteristics of products. Moreover, it was also advised to present consumers with some type of promotional campaign, special offers, discounts, and loyalty points.

Additionally, the novelty, enjoyment, and social characteristics of the modern tools were proposed to be improved by many authors. In this context, it was advised to increase the aesthetics and playful characteristics of metaverse purchasing environments by using bright and dynamic visuals or adding animations, music, fun conversations, or gaming features. Moreover, it was suggested to introduce innovative technologies to convey human quirks such as laughing, sneezes, and fluctuations in tone in the pronunciation of words. The social features were suggested to be improved by integrating informal and natural speeches into greetings and offering causal conversations for simple queries while professional responses for detailed questions. Besides, it was advised to personally name the AI assistants and make them call consumers by names and initiate conversations for a human-like relationship. It was suggested to enable chatbots and robots to express agreement and feelings by using emojis and humor and initiating questions. Specifically, it was advised that the voice bots be equipped with a male voice or elderly people's voice since the elderly and male voices make users feel safe, empathetic, and wise. Finally, it was suggested to equip modern tools with sentiment analysis to understand body language.

Finally, the brands employing virtual tools were recommended to align the offers of the virtual world with those of the real world by putting a greater emphasis on customization. The resemblance of both virtual settings and avatars was strongly suggested. The platforms were suggested to be equipped with a feature enabling consumers to share their virtual experiences in the real-world scene by being able to take a photo. To increase the adoption of these tools, the companies were advised to establish marketing campaigns with authority figures and trusted social media influencers' peers in the collectivist cultures.

4.2 Consumer Approaches Towards Modern Tools in Fashion Retailing

The number of studies concentrating on smart fashion retailing and the modern tools other than AR and VR were lower compared to the smart retailing literature. Nonetheless, similar propositions and approaches were employed to address the motivations and expectations of consumers from these tools.

The smart fashion stores were stated to present many benefits for consumers. Ogunjimi et al. (2021) implied that customer satisfaction and the service quality of brick-and-mortar retail chains could be improved with the employment of innovative technology. The interview-based study established in the UK revealed that the innovative technologies might eliminate the queues in fitting rooms and points of sale and reduce the negative effects of small sizes of fitting rooms. The smart racks and digital signage might enable consumers to find the products, information, and sales points easily. Personalized assistants and self-service technologies could eliminate the need to reach the sales assistant for information. Nonetheless, the authors added that these contributions might not be adopted by the consumers for whom shopping is a social activity. Jain and Gandhi (2021) revealed that AI-based tools could make precise recommendations for products and reduce the purchase duration of the shoppers. Istiqomah and Alfansi (2024) stated that consumers might prefer AI-enhanced fashion e-commerce because of its benefits and ease of use. Kim (2021) put forward that the smart retailing experience, which is a combination of interactivity, benefits, and enjoyment, improved the service quality and satisfaction of consumers irrespective of fashion store type. Finally, Kim et al. (2020) showed that the store atmosphere involving digital elements had a positive impact on consumers' behavioral responses, adding that the consumers of sportswear brands should pay more attention to smart stores.

The modern tools and their effects on fashion retailing were again examined individually, as in the case of smart retailing literature. The chatbots found a broader place in the smart fashion retail literature. Rese et al. (2020) studied the adoption of a text-based chatbot, Emma, in online fashion retailing by German participants and showed that the perceived usefulness, authenticity of conversation, and perceived enjoyment facilitated the adoption of Emma. But the privacy concerns and the immaturity of the technology were considered the main drawbacks of chatbots. Naqvi et al. (2023) implied that entertainment and customization aspects of chatbots were perceived positively, but the accuracy and the credibility of the recommendations were also important. Besides, Mpinganjira et al. (2024) showed that perceived hedonic and utilitarian experiential value had a positive effect, but the perceived social experiential value had a negative effect on shoppers' continued intention to use the chatbot. The authors added that the experiential value could be improved positively by increasing the perceived chatbot anthropomorphism and perceived chatbot intelligence. On the other hand, Zeng et al. (2023) implied that privacy concerns were a challenge for chatbots.

Similar conclusions were also made for the use of service robots in fashion retailing. Song and Kim (2021) showed that self-interest and social interaction were influential on the willingness to share personal information with AI-based robots. In another study, Song and Kim (2022) confirmed that the high service quality perception and adoption of service robots in fashion retailing were associated with usefulness, social capability, and appearance. However, the authors underlined that these findings were valid for consumers who have lower anxiety toward robots.

As mentioned earlier, the smart fashion retail literature broadly discussed the AR-based and VR-based technologies as they present the most convenient features, such as virtual try-ons and virtual visits to stores and fashion shows.

Focusing on the smart fashion mirrors, Alanazi and Alenazi (2023) confirmed the positive impact of perceived usefulness and perceived ease on customer intention and customer satisfaction. Kang et al. (2023) stated that trust in AR applications was a determinant of usage intention, and it can be enhanced by the novelty. Rhee and

Lee (2021) investigated the influence of AR applications on the consumer journey of young Korean female participants and revealed that the consumers with higher product involvement were not as satisfied as the other consumers.

The majority of studies made a comparison between different tools with a major aim of explaining the pros and cons of the adoption, satisfaction, and experience. Concentrating on AR Xue et al. (2023) investigated and compared an AR-branded application with a Magic Mirror working with 42 participants in the UK. The participants indicated that reaching the products and the images via virtual assistants instead of textual information, trying them on virtually, and checking out without waiting for queues were very comfortable and smart. The participants found the AR Branded Application more useful and the Magic Mirror easier to use. Tawira and Ivanov (2023) assessed the virtual try-on applications Forma and Zeekit in Shanghai. In this respect, 61 female participants used Forma for personalization, specifically for the selection of a specific look, and they used Zeekit for customization, with the purpose of updating their wardrobe. It was revealed that photo satisfaction and perceived augmentation were significant features of these tools irrespective of their purpose of use. Besides, fit confidence was stated to be more important for personalization, whereas the Mix-and-Match option was an important component for customization. Nonetheless, the authors implied that the adoption of these tools was highly associated with the satisfaction of the consumers with their bodies. As it turns out, Ivanov et al. (2023) showed that the majority of consumers preferred to work with a preloaded avatar instead of uploading their own photos. Xue et al. (2020) conducted an interview in the UK with the consumers aged between 18 and 34 to understand the VR experience in fashion retailing. The authors stated that hedonically motivated consumers were more open to virtual commerce than utilitarian-motivated consumers. The novelty seekers and younger generation were considered the main adopters of virtual commerce since the older people had difficulty in altering existing shopping modes regarding the handling of the product. Personalized suggestions providing similar types of products, the recommendation of new arrivals, or the matching of items according to the consumers' preferences were favored by the participants. The feature of VR ensuring to do the shopping together with friends and family was assessed positively. Specifically, the virtual and AI-based sales assistants were indicated to present a specific service for the disabled people. However, it was interesting to see that none of the participants became interested in the time efficiency these products provided. Moreover, the main barriers to virtual commerce were listed as the lack of feeling the product in person, being used to traditional shopping, and the expense of additional hardware equipment.

As in the case of the smart retailing literature, the immersion capabilities of both VR-based applications and the metaverse were investigated in detail with the major aim of improving this qualification. Kim et al. (2022) compared a conventional online store with a 360-degree VR store in terms of their ability to enhance the consumer experience by using the flow theory. It was confirmed that the 360-degree VR experience enhanced the flow state and caused an increase in the interest and resulting visit intention towards the store. Lau and Lee (2015) developed a stereoscopic

3D fashion show to understand the immersive and telepresence experiences of consumers. It was found that the immersive experience was enhanced by the spacious environment in the virtual platform. The wide-angel shots, move-along shots, and the lighting effect made the virtual fashion model more alive and enabled consumers to actively investigate the details of products. The try-on experience, as well as the mix-and-match across various clothing, accessories, and even hairstyles, offered a hedonic, game-like experience to consumers. Kim and Choo (2023) compared immersive and non-immersive VR and showed that the perceptual curiosity of the immersive VR was higher. Pillai et al. (2024b) underlined the significance of immersion, interactivity, imagination, and perceived enjoyment in the adoption of the metaverse. However, the authors added that the perceived risk, image barrier, technological anxiety, and traditional barriers deterred metaverse adoption in the mentioned order.

The avatars in the virtual environments were also acknowledged. Arachchi and Samarasinghe (2023) defined the dimensions of an avatar with sensing autonomy, thought autonomy, and action autonomy and showed that all of them improved the novel experience as well as the acceptance. Moreover, Oh et al. (2024) revealed that both avatar–user resemblance and fashion item realism positively affected psychological identification, which led to spending and metaverse usage intention since the individuals, especially the ones who are more sensitive to the environment, could view their avatar as their public self. Concentrating on the anthropomorphism in AR-based m-commerce adoption, Manchanda and Deb (2021) stated that the capacity of AR-based technologies to mimic human traits and intentions made people believe in the innovativeness of the technology and develop positive feelings about it.

On the other hand, the consumer–brand relationship was also addressed in both the VR and metaverse interactions. Agnihotri et al. (2024) revealed that the social presence dimensions (haptic effect, autonomy, immersion, personalization, interactivity, and brand community) had a positive impact on consumer brand engagement and purchase intention of young consumers in the fashion retail metaverse. Yuan et al. (2023) explored the impact of parasocial relationships in virtual reality shopping environments on brand equity with a survey conducted for the UNIQLO brand in China. The findings showed that the technological novelty, physical attractiveness of the celebrity endorser, and social presence improved brand equity positively.

Based on the findings, the following suggestions were made to enhance the adoption of smart fashion retailing. The functionality and acceptance of the modern tools were suggested to be improved by providing up-to-date information, rich content, one-tap payment or self-checkout, and relevant promotions and discounts for those who have purchased earlier by using these tools. Additionally, it was advised to build intimacy by improving the security of the information shared.

Moreover, the resemblance of the virtual world and avatars was advised to be improved. Specifically, it was advised to ensure that the virtual store environment looked like a real store with unique store and interior displays. It was advised to add music in virtual stores. The avatars were suggested to be created in human-like formats with increased sensing autonomy by using sophisticated facial scanning platforms, natural language processing algorithms, and sight recognition and location features.

Finally, the entertainment and social features were suggested to be improved. Specifically, it was suggested to brand managers to present some videos on products, trends, and styles; create online games; build digital avatar groups; and arrange events that bring customers together. It was advised to work with physically attractive and beloved celebrities, who are accepted as friends by the users.

4.3 Consumer Characteristics and Adoption of Smart Retailing

Since modern tools and smart applications are comparatively new topics, consumer characteristics were underexplored in terms of their impact on consumer behavior literature.

In these studies, the innovativeness of consumers and being experienced users were stated to have a more positive impact on consumer behavior. Fazal-e-Hasan et al. (2021) implied that the extent and degree to which the individual is exposed to technological innovation was influential on consumer intention. The innovative consumers were stated to have more positive experiences as they might engage with the smart retail technology in both cognitive and emotional manners (Adapa et al., 2020). Moreover, Chang and Chen (2021) showed that consumers with a higher level of technology readiness would obtain more utilitarian benefits in the smart shops. On the contrary, the consumers were stated to hesitate and be skeptical and avoid using technologies if they were not technology-savvy (Soon et al., 2020). The impact of past experience was also shown to influence the attitude and intention towards a specific type of technology (Fazal-e-Hasan et al., 2021). Kim et al. (2021) revealed that a highly experienced group had a higher level of perceived usefulness and perceived enjoyment regardless of their interactivity level. Similarly, Lin (2022) confirmed that the experienced consumers' attitudes prioritized the convenient features of modern tools over their novelty.

Some researchers concentrated on the impact of sociodemographic characteristics of consumers on their behavior. Saprikis et al. (2020b) showed that males and females were both influenced by the performance expectancy and the social influence, but the females were much more influenced by the social influence. Additionally, Rese et al. (2020) implied that females approached the modern tools from a hedonic perspective, whereas males evaluated them from a utilitarian perspective.

The studies also revealed that the national culture plays a pivotal role in determining attitudes. Concentrating on smart shopping carts, Roy et al. (2020) revealed that the perceived flow had a stronger influence on the Australian sample and the retailer support and perceived attractiveness had a stronger influence on the Indian sample. In a study where a comparison was made between German and Italian samples, it was found that there was a positive association between ease of use and users' attitudes toward the adoption of augmented reality systems for only the German sample (Teles Roxo & Quelhas Brito, 2020). Having developed a model for customer interactions with smart technologies in shopping malls in the United Kingdom and the United Arab Emirates, Ameen et al. (2022b) revealed trust, interface design, and relationship commitment were important for both samples, but consumer peer interaction was only important for the United Kingdom sample. On the other hand, the economic value (resale potential) was found to have a stronger influence on NFT collection for U.S. consumers than for South Korean consumers (Balakrishnan et al., 2024).

Among the sociodemographic characteristics, especially age and generation, was assessed in many studies. The technology and perceived usability were declared to be influenced by age because of the lower skills and higher technology anxiety of elderly consumers (Manis & Choi, 2019). Manis and Choi (2019) expressed that younger consumers valued the usefulness of technology more, whereas older consumers valued perceived ease of use. Moreover, the arousal because of the media richness was confirmed to decrease with age (De Amorim et al., 2022). Ferber and Vaziri (2024) showed that the younger participants became more satisfied with human–robot interactions than the older participants.

The reaction of especially Generation Y and X drew particular interest from researchers. Arachchi and Samarasinghe (2024) conducted a survey among Generation X and Y in Sri Lanka and India and found that Generation Y was more innovative in adopting speech recognition applications. Sharma et al. (2024) confirmed the most important attributes of digital assistants as perceived interactivity, perceived anthropomorphism, perceived social presence, and information quality for millennials. On the other hand, it was revealed that Generation Z was stated to possess unique characteristics that set them apart from baby boomers, Generation X, and millennials (Pfeifer et al., 2023). Generation Z, also called the I-generation, net-gen, or digital natives, are recognized as highly educated, technologically savvy, innovative, and creative. Therefore, Generation Z was stated to have high interest in new technologies and ease of using these technologies and feel safe while using them (Pfeifer et al., 2023; Ng et al., 2021). Ng et al. (2021) investigated the expectations of Generation Z consumers regarding smart retail technology in Malaysia and found out that Generation Z consumers valued functionality and entertainment features most. Guo and Luo (2023) implied that Generation Z consumers would feel more emotionally supported if personalization of the modern tools were provided and a conversational tone was employed.

The findings of a limited number of studies put forward the significance of the consumer characteristics in their motivations and expectations, implying the requirement of considering these factors when developing strategies for the implementation of these tools. Furthermore, the findings highlighted the substantial importance of experience for the adoption of these modern tools.

5 Empirical Study

The literature review on consumer behavior showed that there were numerous studies in literature concentrated on either the smart stores or the individual modern tools. Although the modern tools are used together with the traditional tools in mobile fashion stores, the consumer approaches towards the traditional tools and modern tools were not considered together in any of these studies. Because of this gap in the literature, this study aims to analyze the service quality perception of both modern tools (including the smart assistants and virtual try-ons) and traditional tools (servicescape in mobile shopping) in the enhancement of store atmospherics and experience in mobile shopping applications for fashion brands. Since the impact of sociodemographic characteristics has been considered in a limited number of studies, the secondary aim of this study is to investigate the impact of sociodemographic characteristics such as gender, age, nationality, and different levels of experience on consumer behavior.

5.1 Research Model and Data Analysis

The model applied in the study is the combination of the SOR algorithm and the UTAUT model. The SOR algorithm was preferred because of its superiority in analyzing consumer reactions in a specific environment. The original SOR paradigm involves three major variables: stimuli, organism, and response. The stimuli variables refer to the inputs inciting or increasing an action; organism variables refer to the emotional intervening mechanism; and response variables refer to outcomes derived from the intervening mechanism. As the original SOR paradigm was criticized for lacking the cognitive aspects, and recent studies confirmed that behavioral patterns are shaped by cognition rather than emotion (Saricam, 2023), the model employed in this study used the SOR algorithm as a framework, including both cognitive and affective evaluation as organism constructs. The stimuli variables were selected as the service quality perception about modern tools and traditional tools. The UTAUT model was employed to measure the service quality perception about modern tools in different aspects, namely performance expectancy, effort expectancy, social influence, and facilitating conditions. The service quality perception about the traditional tools was measured in three elements, namely design cue, navigation cue, and information cue.

On the other hand, since the consumers' characteristics associated with age, gender, and nationality were stated to act as external variables in evaluating and adopting the technology (Plotkina et al., 2022), and the previous technological experience had a significant positive effect on consumer behavior (Kim et al., 2021), the gender, age, nationality, and experience level were taken as control variables in this study.

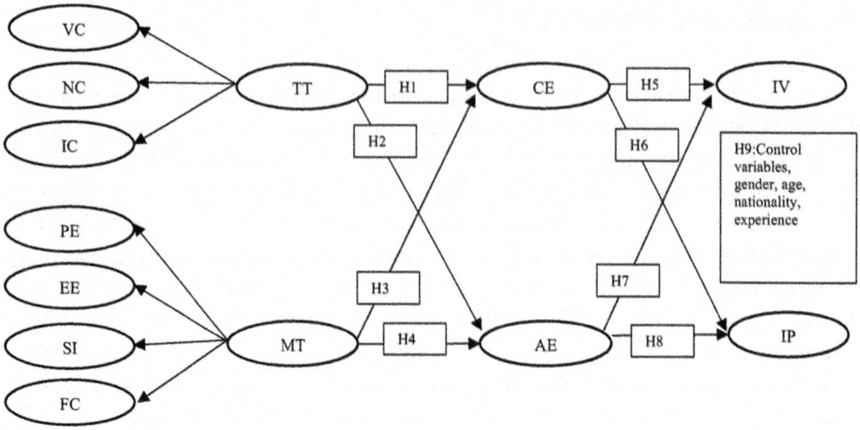

Fig. 2 The conceptual model

Based on all these, the conceptual model proposed by this study is given in Fig. 2 with an explanation of the hypotheses and the variables below. In the conceptual model, the service quality perception in regard to traditional tools (TT) and modern tools (MT) was considered a second-order construct. In the conceptual model, visual cue (VC) refers to the service quality perception about the attributes provided by traditional tools that can stimulate human vision. Navigational cue (NC) corresponds to the service quality perception related to diverse functions of traditional tools such as navigation bars, search options, and logical menu hierarchies provided by traditional tools to guide users. Information cue (IC) is related to the service quality perception of information delivered via diverse formats by traditional tools in mobile stores. Performance expectancy (PE) refers to the service quality perception of the modern tools in regard to the benefits of performing certain activities. Effort expectancy (EE) is associated with the service quality perception of the modern tools related to the ease of learning these tools. Social influence (SI) refers to service quality perceptions of the modern tools about others' thoughts and feelings. Facilitating conditions (FC) are associated with the service quality perception in regard to the knowledge and resources required to use the modern tools. Furthermore, cognitive evaluation (CE) defines the degree to which a person is interested in interacting with the tools in question, whereas affective evaluation (AE) defines the degree to which a person feels passionately about interacting with the tools in question. Finally, intention to visit (IV) refers to the willingness to visit the mobile platforms now and in the near future, and intention to purchase (IP) refers to the willingness to visit the mobile platforms now and in the near future.

14 questions for VC, NC, and IC were derived from the study established by Koo and Park (2017); 16 questions for PE, EE, SI, and FC were derived from the study established by Rahman et al. (2020); 8 questions for CE and AE and 8 questions for

IV and IP were derived from the studies by Saricam (2023) and Manis and Choi (2019).

In the conceptual model;

H_1: The service quality perception of traditional tools has a positive impact on cognitive evaluation.
H_2: The service quality perception of traditional tools has a positive impact on affective evaluation.
H_3: The service quality perception of modern tools has a positive impact on cognitive evaluation.
H_4: The service quality perception of modern tools has a positive impact on affective evaluation.
H_5: The cognitive evaluation has a positive impact on intention to visit mobile stores.
H_6: The cognitive evaluation has a positive impact on intention to purchase from mobile stores.
H_7: The affective evaluation has a positive impact on intention to visit mobile stores.
H_8: The affective evaluation has a positive impact on intention to purchase from mobile stores.
H_9: The consumer characteristics and experience level are the control variables.

The data was collected via an online survey conducted with a target of reaching at least 500 participants. A questionnaire was developed for the survey involving two parts. The first part included the questions about the sociodemographic characteristics of the participants, such as gender, age, and nationality, and questions regarding the previous experience of the consumers with the modern tools used in mobile application tools. The second part included questions regarding the service quality perception of traditional and modern tools in mobile application tools, as well as the affective and cognitive evaluation of these tools and the behavioral intentions, specifically the intention to visit and the intention to purchase. A five-point Likert scale was used in the questions in the second part. The data set was analyzed by using IBM SPSS 27 and IBM AMOS 22 statistical programs. The univariate normality was examined with the skewness and kurtosis values. The factors were confirmed by using confirmatory factor analysis (CFA). The modification indices in error terms of direct variables were examined to improve the model fit. The factor structure was evaluated in terms of internal consistency, composite reliability, and divergent validity. The hypotheses were tested, and the strength of the relationships between factors was derived by structural equation modeling (SEM). The predictive ability of the model was evaluated with the assessment of squared multiple correlation values (SMC) for the dependent variables. A multigroup invariance test was performed to assess the equivalence in the measurement model between the groups and compare the responses of consumer groups in regard to gender, age, nationality, and experience level.

5.2 *Results and Discussion*

The data was collected from 623 participants. The demographic profile as well as the data in regard to the experience level of the consumers was given in Table 1.

The number of female participants was slightly higher than male participants, where 43.8% of participants were composed of males, and 56.2% were composed of females. The majority of the participants (64.4%) were aged below or equal to 25. Besides, the majority of the respondents (69.3%) were Turkish citizens, while the rest were composed of people belonging to 12 different nationalities. The participants, who experienced the modern tools in either apparel retail websites and mobile applications or in any other type of retail company, were considered as the consumers who have experienced the modern tools. In this respect, 53.3% of the participants have experienced modern tools.

The factor structure was confirmed by CFA based on maximum likelihood estimation. The initial goodness-of-fit values were given in the third column of Table 2. Three questions—one question from cognitive evaluation, one question from affective evaluation, and one question from intention to visit—were deleted because

Table 1 Demographic profile of the sample

Parameters	Group	Frequency	%	Parameters	Group	Frequency	%
Gender	Male	273	43.8	Nation	Non-Turkish citizen	191	30.7
	Female	350	56.2		Turkish citizen	432	69.3
Age	Below or equal to 25	401	64.4	Experience level	No experience	291	46.7
	Above 25	222	35.6		Have experience	332	53.3

Table 2 The goodness of fit indices

The goodness of fit indices	Reference values for comparison	Initial CFA	Final CFA	Initial SEM	Final SEM
Chi-Square		2687.901	1860.556	2212.085	2199.326
Degree of freedom		967	827	831	830
Chi-square/degree of freedom	≤3	2.780	2.250	2.662	2.650
Goodness of fit index (GFI)	≥0.9	0.834	0.874	0.854	0.855
Adjusted goodness of fit index (AGFI)	≥0.8	0.814	0.856	0.834	0.835
Comparative fit index (CFI)	≥0.9	0.957	0.972	0.962	0.963
Incremental fit index (IFI)	≥0.9	0.957	0.972	0.962	0.963
Normed fit index (NFI)	≥0.9	0.934	0.951	0.941	0.942
Root mean square of approximation (RMSEA)	≤0.08	0.053	0.045	0.052	0.052
Root mean square residual (RMR)	≤0.08	0.051	0.046	0.059	0.059

of lower factor loadings. The modification indices for error terms were checked, and the error terms having high shared variance were correlated to improve the model fit. The improved goodness-of-fit values were given in the fourth column of Table 2.

The internal consistency of the factors was evaluated by using Cronbach's alpha values shown in Table 3.

The Cronbach's alpha values were between 0.942 and 0.979, well above 0.6, implying internal consistency. The convergent validity was confirmed as the composite validity values were between 0.943 and 0.972 and higher than 0.6, and; the average variance extracted values were between 0.845 and 0.0898 and above 0.5. The divergent validity was also confirmed according to Fornell and Larcker's criterion, as the correlation between the factors was lower than the square root of the average variance extracted values of each factor in concern.

The structural equation model was used to test the hypotheses within the 0.05 level of significance, and it was confirmed that all the hypotheses were found to be significant. The strength of the relationships was given in Fig. 3. The SMC values for the dependent variables were shown in the blocks of dependent variables.

The findings showed the service quality perception of both modern tools and traditional tools was influential on both cognitive evaluation and affective evaluation. Moreover, the service quality perception of modern tools was found to have a higher impact on both cognitive evaluation and affective evaluation compared to the

Table 3 Convergent and discriminant validity analysis

	CA	CR	AVE	TT	MT	CE	AE	IV	IP
TT	0.977	0.958	0.884	**0.940**					
MT	0.979	0.972	0.898	0.832	**0.948**				
CE	0.949	0.950	0.864	0.817	0.871	**0.930**			
AE	0.949	0.949	0.861	0.801	0.855	0.849	**0.928**		
IV	0.942	0.943	0.845	0.715	0.792	0.808	0.788	**0.919**	
IP	0.956	0.957	0.847	0.663	0.776	0.758	0.776	0.909	**0.921**

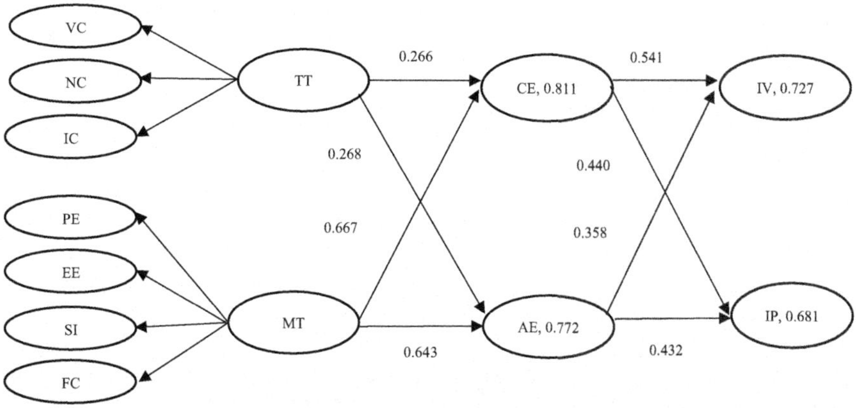

Fig. 3 The significant relationships and the strength of the relationships

service quality perception of traditional tools. Whereas the relationship between the service quality perception of traditional tools and cognitive evaluation is almost equal to the relationship between the quality perception of traditional tools and affective evaluation, the service quality perception of modern tools was shown to have a higher impact on cognitive evaluation than on affective evaluation.

Moreover, the impact of cognitive evaluation on both intention to purchase and intention to visit was shown to be higher compared to that of affective evaluation. Whereas the influence of cognitive evaluation on intention to visit was higher than intention to purchase, it was vice versa for the influence of affective evaluation. The predictive ability of the model also revealed that the model can explain cognitive evaluation better than affective evaluation. Similarly, a higher amount of variance was explained for the intention to visit.

The multigroup analyses were established after the metric invariance test. The invariance tests resulted in nonsignificant p-values, revealing that the groups were not different at the model level. The regression weights for different groups were given in Figs. 4, 5, 6 and 7, and the squared multiple correlation values for the dependent variables were given in figure captions.

When the results of the multigroup analysis based on gender shown in Fig. 4 were analyzed, it was seen that the impact of the service quality perception of modern tools was higher on both cognitive evaluation and affective evaluation for both genders. However, the comparison of genders revealed that the impact of the service quality perception of modern tools on cognitive evaluation was found to be higher for males. Besides, the cognitive evaluation was found to be more influential on the intention to purchase than the intention to visit for males, whereas the affective evaluation had a higher impact on the intention to purchase for females. The model's predictive ability was almost the same for both genders.

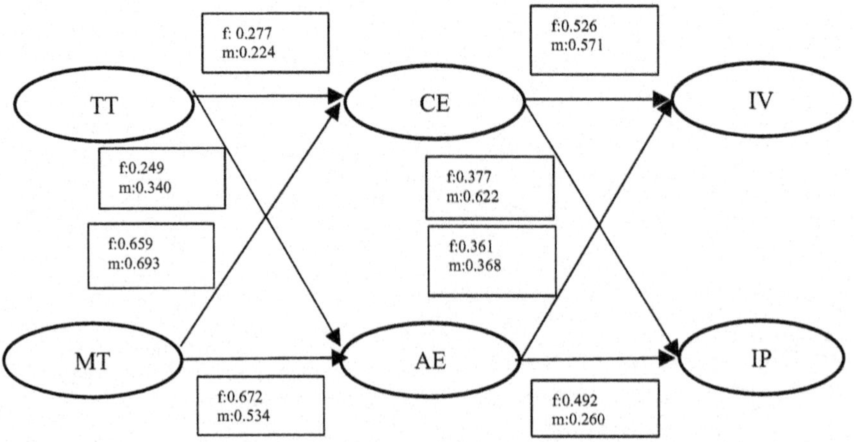

Fig. 4 The model results according to the gender, f: female, m:male (SMC: CE:f = 0.813,m = 0.798; AE:f = 0.790, m = 0.714; IV:f = 0.711,m = 0.778; IP:f = 0.681,m = 0.698)

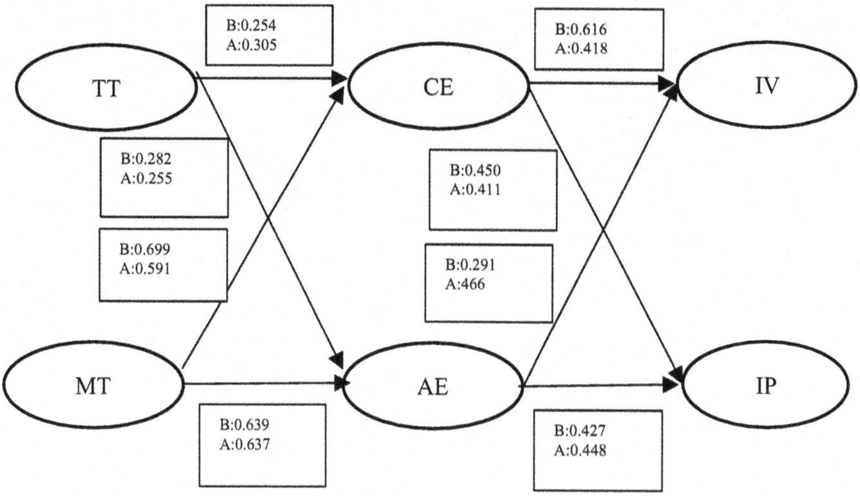

Fig. 5 The model results according to the age, B:below or equal 25, A:above 25 (SMC: CE:B = 0.850, A = 0.737; AE:B = 0.790, A = 0.737; IV:B = 0.757, A = 0.578; IP:B = 0.700, A = 0.641)

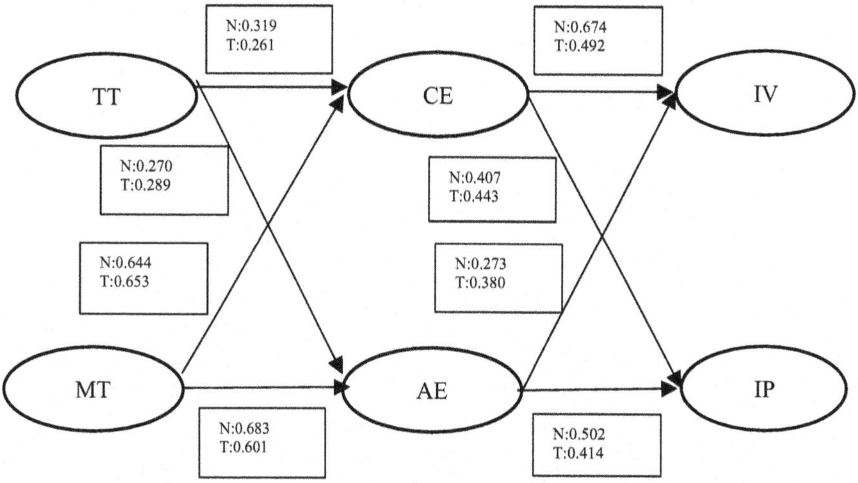

Fig. 6 The model results according to nationality, N:non-Turkish, T:Turkish, (SMC: CE:N = 0.873,T = 0.770; AE:N = 0.859, T = 0.727; IV:N = 0.848, T = 0.665; IP:N = 0.772, T = 0.642)

The multigroup analysis based on age was established for the two age groups, namely, "below or equal to 25" and "above 25," as shown in Fig. 5.

When the results of the multigroup analysis based on age were analyzed, it was seen that the impact of the service quality perception of modern tools was higher than that of traditional tools on both cognitive evaluation and affective evaluation

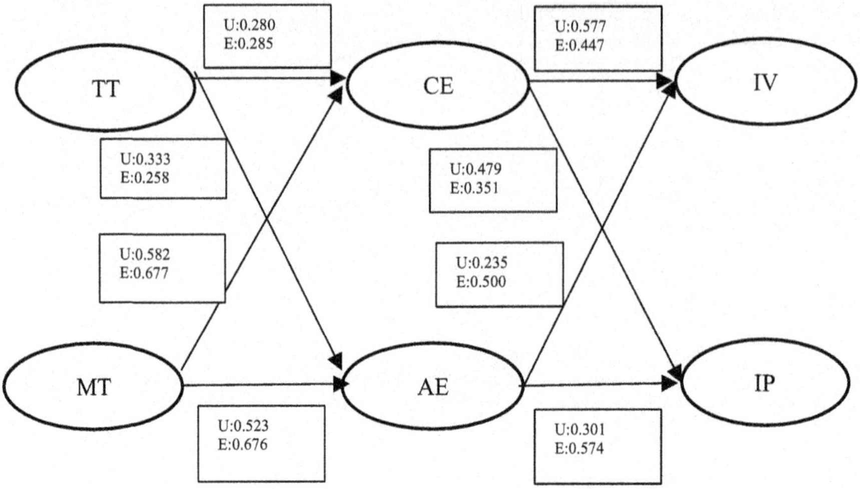

Fig. 7 The model results according to experience level, U:Unexperienced, E:Experienced (SMC: CE:U = 0.657, E = 0.876; AE:U = 0.640, E = 0.830; IV:U = 0.563, E = 0.831; IP:U = 0.778, E = 0.796)

for both groups. For the consumer group below or equal to 25 years old, the service quality perception of modern tools had a higher impact on cognitive evaluation than on affective evaluation, whereas the service quality perception of traditional tools was stated to have a higher impact on affective evaluation than on cognitive evaluation. On the contrary, for the consumer group aged above 25, the impact of the service quality perception of modern tools was stated to be higher on affective evaluation than on cognitive evaluation, whereas the impact of the service quality perception of traditional tools on cognitive evaluation was higher than the impact on affective evaluation. Moreover, the consumer groups aged below and equal to 25 think that the impact of cognitive evaluation on both intention to visit and intention to purchase was higher than the impact of affective evaluation. On the other hand, there was a higher impact of affective evaluation on both intention to purchase and intention to visit for the consumer groups aged above 25. The model's predictive ability was higher for the consumer group aged below or equal to 25.

The multigroup analysis based on nationality was established for the two consumer groups whose results are shown in Fig. 6.

Whereas the first group of consumers was composed of non-Turkish citizens, the second consumer group was composed of Turkish citizens. For both groups, the impact of the service quality perception of modern tools was higher than that of traditional tools on both cognitive and affective evaluation. Even though there was a slight difference, the impact of the service quality perception of modern tools was higher on cognitive evaluation than on affective evaluation for Turkish citizens. On the other hand, the cognitive evaluation had a higher impact on the intention to visit, and the affective evaluation had a higher impact on the intention to visit for non-Turkish citizens, whereas the cognitive evaluation was found to have a higher impact on both the intention to visit and the intention to purchase for Turkish citizens. The model's predictive ability can be considered to be higher for non-Turkish citizens.

The multigroup analysis was also established for the experienced and unexperienced consumer groups, as shown in Fig. 7. The experienced consumers were composed of the consumers who had previously experienced modern tools in mobile applications either for apparel products or for any other type of products. The impact of the service quality perception of modern tools was found to be higher than that of traditional tools both on cognitive evaluation and on affective evaluation for both consumer groups. On the other hand, the impact of the service quality perception of modern tools was higher on cognitive evaluation than on affective evaluation for the inexperienced group, whereas these impacts were nearly the same for experienced groups. Moreover, the impact of cognitive evaluation was higher on both intention to visit and intention to purchase for the inexperienced group, whereas the impact of affective evaluation was higher for the experienced group of consumers. The model's predictive ability can be considered to be higher for the experienced consumers.

The results revealed that the service quality perception of traditional tools has an almost equal impact on both cognitive and affective evaluation in this study. The relationship between the traditional tools and cognitions was already confirmed in the previous studies. Specifically, Kang et al. (2020) revealed that the realistic graphics allowed the products to be inspected in greater detail by increasing the perceived informativeness, adding that the visual-spatial cue had a positive impact on cognitive evaluation. Moreover, the intuitive, convenient, and logical navigational cues were associated with cognitive evaluation by Floh and Madlberger (2013) and Hsieh et al. (2014). On the other hand, the association of the traditional tools with emotions was also confirmed in previous studies. For instance, interesting graphic features or product images were stated to create pleasure (Kim & Lennon, 2010). The website design elements were underlined to positively affect consumer arousal (Kim et al., 2015), and the availability of detailed information was found to have a positive effect on emotions (Floh & Madlberger, 2013; Gao & Bai, 2014), which results in pleasure and the feeling of being in control.

The service quality perception of modern tools was also found to have a positive impact on both cognitive and affective evaluation in this study. In fact, the capabilities of the modern tools and the less effort required to use these tools without needing further control were mentioned in previous studies. Specifically, Huang and Qin (2011) and Alanazi and Alenazi (2023) emphasized their impact on the accurate evaluation of the apparel fit and size in less time. Besides, Chang et al. (2023) and Lin (2022) drew attention to the increased control of operations, which might make consumers feel good about themselves and the modern tools.

Furthermore, service quality perception of modern tools about social influence and facilitating conditions was also associated with both cognitive and affective evaluation in this study. These points also find support from the literature. In fact, Lin (2022) confirmed the positive impact of social influence on attitudes toward intelligent unmanned convenience stores. Shaw and Sergueeva (2019) expressed that the higher tendency to use modern tools is associated with the wish to impress others by showing them that they are using new technology. In regard to the positive impact of facilitating conditions, Roy et al. (2020) proposed that the availability of support during customer interaction with smart technologies could reduce anxiety, which might result in a favorable cognitive and affective evaluation of modern tools.

On the other hand, the service provided by modern tools was found to be more influential than that of traditional tools on both cognitive evaluation and affective evaluation in this study. This might be attributed to the fact that the modern tools provided more information and resulted in a more enjoyable experience (Ettis, 2017). In addition, the service provided by modern tools has a higher impact on cognitive evaluation than affective evaluation in this study. This is quite expected, as customer–product interactions are more intelligent due to modern tools. In fact, virtual reality is more informative than a still image (Kang et al., 2020). The websites using human models to show how garments should fit or employing technological devices such as color swapping on clothing and enlargement of product images might give shoppers a better idea of how garments will look (Kim & Lennon, 2010). Moreover, Perumal et al. (2022) and Oyman et al. (2022) implied that the features related to the usefulness of these tools were more important than the entertaining characteristics of these tools.

This study also revealed that both cognitive evaluation and affective evaluations were influential on intention to visit and purchase. These findings were also parallel with the previous findings. Specifically, Kang et al. (2020) also confirmed that cognitive evaluation has a positive impact on purchase intention, and Soon et al. (2020) revealed that emotions mediated the relationship between exposure to the applications and attitude towards using them. Besides, Chang et al. (2023) implied that the hedonic motivations, associated with the feelings, and utilitarian motivations, associated with the thoughts, had driven consumer behavior toward smart stores.

Furthermore, this study showed that the cognitive evaluation had a slightly higher impact on intention to visit, whereas affective evaluation had a higher impact on intention to purchase. This finding might be attributed to the fact that mobile applications, especially those integrating modern tools, increase the curiosity of the participants to visit the store, and the participants deploy some feelings before making the purchase of fashion products.

The multigroup analysis in terms of gender revealed that the impact of the service quality perception of modern tools on cognitive evaluation and the impact of cognitive evaluation on purchase intention were higher for males. The results may be due to the fact that the males are more concentrated on utilitarian perspectives (Rese et al., 2020) and therefore cognitions.

The age comparisons revealed that the service quality perception of modern tools was more related to cognitive evaluation, which in turn resulted in the higher tendency for both intention to visit and intention to purchase for younger people. The reason for this is that the age group of below 25 involves more participants from Generation Z consumers. Indeed, it is known that Generation Z is accustomed to smart technologies, possesses a great deal of knowledge, and shows more interest in emerging technologies (Ng et al., 2021). Moreover, the new technologies are considered service enhancements by Generation Z (Adapa et al., 2020).

In this study, the service quality perception of modern tools has a slightly higher impact on cognitive evaluation, which is the main determinant for both intention to visit and intention to purchase for Turkish people. These findings might be attributed to the lifestyle differences and the ratio of smartphone users in Turkey. Since the utilization rate of smartphones is higher compared to other developing countries (Ince & Kılıç, 2016), this might enable Turkish citizens to have a higher tendency towards modern tools.

Finally, the group comparison in terms of experience in this study revealed that the experienced consumers are more motivated to deploy affective evaluation for the modern tools, which resulted in more willingness to both intend to visit and purchase. This finding is actually in parallel with the previous findings. The positive impact of the experience was validated in the customer's inclination toward online purchases in previous studies (Manis & Choi, 2019; Miao et al., 2022). Moreover, it was expressed that experienced consumers were more familiar with the procedures and had lower barriers to accessing these services (Lin, 2022).

5.3 *Practical Implications and Recommendations*

The study has both theoretical and practical contributions. Theoretically, the first contribution of this study lies in examining the service quality perception of both modern tools and traditional tools at the same time. This allowed us to make a comparison of the effect of those tools on behavioral intention. Secondly, the conceptual model combined the SOR framework with the UTAUT model, and the model investigated both the cognitive aspects and affective aspects of the behavioral intentions. Even though the UTAUT model is one of the most widely used models to identify technology acceptance and motivations, it has not been employed in any study with the SOR algorithm, which has the superiority to investigate the reactions of the consumers towards the external stimuli. Moreover, the cognitive evaluation was not established in any studies employing a similar framework until now. Last but not least, comparisons were made for different consumer groups and different experience levels. Even though some studies made comparisons in terms of the characteristics of consumers, there is no study covering the comparisons in terms of gender, age, nationality, and experience at the same time.

The findings of this study revealed that modern tools are favored by all the consumers irrespective of the sociodemographic characteristics as well as the experience level of the consumers. The findings provide clear evidence that modern tools in concern (smart assistants and virtual try-ons) are certain to attract the attention of consumers. Nonetheless, it is understood that the features of these tools were highly concentrated on informativeness since the modern tools were mostly associated with cognitive evaluation. Whereas this feature can be taken for granted by some consumers, it is not an appealing property for every consumer.

Since the feature of modern tools to provide much information is attractive to male consumers, for whom the utilitarian benefits were more important, and younger consumers, who are more technology savvy, it is advised to increase the detail level of the information as well as the recency of it that are provided by the modern tools. Specifically, much information is suggested to be provided about the origins, content, and care of the fashion products in product recommendations. Moreover, it would be better to make product recommendations according to past interactions, fashion trends, and the specific event for which the product is intended to be worn. On the other hand, the tools are advised to have a direct connection to self-checkout systems and alternative payment options, such as payment with cryptocurrency or via face recognition.

Considering that the young generation, specifically Generation Z, is also interested in the entertaining properties of the service they are getting, it can be advised for the brands concentrating on this group of consumers to provide mix-and-match options within the product recommendation and present enhanced animations and/or try-ons that showcase the fit and appearance of the product during different body movements.

As the females and older consumers were found to have a tendency to associate with modern tools via affective evaluations, it is advised that the brands concentrating on these groups of consumers trigger emotions. Specifically, it can be advised to improve the brands to integrate social presence into the modern tools and improve personalization. The social aspects of the modern tools can be enhanced with the additional features based on the sentiment analysis, face recognition, and so on. Since these consumers prefer to approach the smart assistant as they approach their friends, the integrated face recognition feature might enable the tools to recognize consumers, resulting in the development of an emotional bond. The friendly talks with the smart assistant, including the valedictions, greetings, and personal sharing, might enhance the adoption of these tools. Moreover, it might be suggested to implement a social sharing feature into the modern tools, which allows consumers to be able to contact their friends and families. Additionally, some external social contribution is also advised, such as the promotion of the modern tools via some celebrities and influencers.

Since the other finding of this study revealed that consumers who tried smart retailing technology before have a more positive approach to modern tools, it would be strategic and valuable to convince consumers to experience these tools. The mobile shopping platforms are for individual use and therefore related to habits, however. Therefore, it is advised that the brands and companies first make consumers aware of these tools in the physical platforms. This means that the introduction of these tools is strictly suggested first in the physical stores, where the sales personnel also assume responsibility for teaching the use of these tools. In this way the consumers feel more willing to use these tools in their personal use.

Even though the study contributes to literature theoretically and practically as given above, it is not free of limitation. The nationality was observed between two groups of consumers composed of Turkish and non-Turkish citizens. The non-Turkish citizens covered 12 different nationalities to get further insight. It would be better to work with a sample of more homogenous groups. Besides, the modern tools in this study cover any type of modern tool, including smart assistants and virtual try-ons. The application of the study concentrating on a specific group of modern tools might be beneficial to reveal more detailed and specific conclusions.

Acknowledgement The authors would like to thank to Ayse Sena DAL, Emine Zeynep SISMAN, Arif Berk DEMIREZER, Zaid JDAYEH, Aigerim YERBOLGANOVA and Ibrahim FOSTOK for their contribution in conducting the survey of empirical study as a part of their Senior Design Project.

References

Adapa, S., Fazal-e-Hasan, S. M., Makam, S. B., Azeem, M. M., & Mortimer, G. (2020). Examining the antecedents and consequences of perceived shopping value through smart retail technology. *Journal of Retailing and Consumer Services, 52*, 101901.

Adegeest, D. (2023, August 24). *Ralph Lauren launches holiday-themed experience on Roblox*. FashionUnited. https://fashionunited.com/news/fashion/ralph-lauren-launches-holiday-themed-experience-on-roblox/2021120944371.

Afolabi, O. (2023, May 7). *7 companies with Metaverse stores—But are they worth visiting?* MUO. https://www.makeuseof.com/companies-with-metaverse-stores

Agnihotri, D., Chaturvedi, P., Swarup, K., Mathur, A., Tripathi, V., & Singh, N. (2024). Does social presence drive customer brand engagement and purchase intention in the fashion retail metaverse? The moderating role of self-efficacy. *Internet Research*.

Ahn, J., Kim, J., & Sung, Y. (2022). The effect of gender stereotypes on artificial intelligence recommendations. *Journal of Business Research, 141*, 50–59.

Aiolfi, S. (2023). How shopping habits change with artificial intelligence: Smart speakers' usage intention. *International Journal of Retail & Distribution Management, 51*(9/10), 1288–1312.

Akdim, K., & Casaló, L. V. (2023). Perceived value of AI-based recommendations service: The case of voice assistants. *Service Business, 17*(1), 81–112.

Alam, S., Abdullah, S., Kokash, H., Ahmed, S., & Omar, N. (2024). Consumer's acceptance of retail service robots: Mediating role of pleasure and arousal. *Journal of Decision Systems*, 1–27.

Alanazi, T. M., & Alenazi, S. A. (2023). The impact of using smart fashion mirrors on perceived customer satisfaction in fashion retailing stores. *Information Science Letters An International Journal, 12*(6), 2401–2409.

Albăstroiu, I., & Felea, M. (2015). Enhancing the shopping experience through QR codes: The perspective of the Romanian users. *Amfiteatru Economic Journal, 17*(39), 553–566.

Albrecht, C. M., & Honekamp, M. (2022). Humanoid robots in retail: Whether and how do consumers perceive deep and surface acting of robots in customer-robot interactions?. In *International conference marketing trends, Venice, Italy*, available at: http://archives. marketing-trends-congress. com/2022/pages/PDF/076. pdf (accessed 2 Sept. 2023).

Alexander, B., & Varley, R. (2025). Retail futures: Customer experience, phygital retailing, and the Experiential Retail Territories perspective. *Journal of Retailing and Consumer Services, 82*, 104108.

Álvarez Márquez, J. O., & Ziegler, J. (2020). In-store augmented reality-enabled product comparison and recommendation. In *Proceedings of the 14th ACM conference on recommender systems* (pp. 180–189).

Ameen, N., Cheah, J. H., & Kumar, S. (2022a). It's all part of the customer journey: The impact of augmented reality, chatbots, and social media on the body image and self-esteem of generation Z female consumers. *Psychology & Marketing, 39*(11), 2110–2129.

Ameen, N., Hosany, S., & Paul, J. (2022b). The personalisation-privacy paradox: Consumer interaction with smart technologies and shopping mall loyalty. *Computers in Human Behavior, 126*, 106976.

Arachchi, H. D. M., & Samarasinghe, G. D. (2023). Impact of artificial autonomy avatar on consumer acceptance of fashion retail services. In *2023 international research conference on smart computing and systems engineering (SCSE)* (Vol. 6, pp. 1–8). IEEE.

Arachchi, H. D. M., & Samarasinghe, G. D. (2024). Impact of embedded AI mobile smart speech recognition on consumer attitudes towards AI and purchase intention across Generations X and Y. *European Journal of Management Studies, 29*(1), 3–29.

Araújo, T., & Casais, B. (2020). Customer acceptance of shopping-assistant Chatbots. In *Marketing and smart technologies. smart innovation, systems and technologies* (pp. 278–287). Springer.

Arora, M., Goel, R., & Singla, J. (2024). Innovation or intrusion? Examining employee willingness to collaborate with AI-powered service robots in retail. In *Augmenting Retail Reality, Part A: Blockchain, AR, VR, and the Internet of Things* (pp. 97–116). Emerald Group Publishing.

Arthur, R. (2016, September 11). *Tommy Hilfiger launches chatbot on Facebook Messenger to tie to Gigi Hadid collection*. Forbes. https://www.forbes.com/sites/rachelarthur/2016/09/11/tommy-hilfiger-launches-chatbot-on-facebook-messenger-to-tie-to-gigi-hadid-collection/

Arthur, R. (2017, August 10). *Fashion site ASOS launches visual search tool to aid inspiration and discovery for shoppers*. Forbes. https://www.forbes.com/sites/rachelarthur/2017/08/10/asos-visual-search/

Balakrishnan, J., Das, R., Alalwan, A. A., Raman, R., & Dwivedi, Y. K. (2024). Informative and peripheral metaverse: Which leads to experience? An investigation from the viewpoint of self-concept. *Computers in Human Behavior, 156*, 108223.

Balchandani, A., Rölkens, F., & D'Auria, G. (2024, November 11). *The state of fashion 2025: Challenges at every turn*. McKinsey & Company. https://www.mckinsey.com/industries/retail/our-insights/state-of-fashion

Baskoro, H. A., & Amini, A. (2020). Analysis of factors influencing consumer intention toward Indonesia QR mobile payment. In *Understanding Digital Industry* (pp. 112–116). Routledge.

Bousquette, I. (2025, January 27). *How tech helped Levi's ride the 'baggy jeans' trend*. Wall Street Journal. https://www.wsj.com/articles/how-tech-helped-levis-ride-the-baggy-jeans-trend-f290721d

Business Today Desk. (2022, June 17). *After Gucci, TAG Heuer, French fashion major Balenciaga will also accept crypto payments*. Business Today. https://www.businesstoday.in/crypto/story/after-gucci-tag-heuer-french-fashion-major-balenciaga-will-also-accept-crypto-payments-334719-2022-05-23

Caro, F., & Sadr, R. (2019). The Internet of Things (IoT) in retail: Bridging supply and demand. *Business Horizons, 62*(1), 47–54.

Chakraborty, D., Polisetty, A., & Rana, N. P. (2024). Consumers' continuance intention towards metaverse-based virtual stores: A multi-study perspective. *Technological Forecasting and Social Change, 203*, 123405.

Chang, V., Chen, W., Xu, Q. A., & Xiong, C. (2021). Towards the customers' intention to use QR codes in mobile payments. *Journal of Global Information Management (JGIM), 29*(6), 1–21.

Chang, Y. W., & Chen, J. (2021). What motivates customers to shop in smart shops? The impacts of smart technology and technology readiness. *Journal of Retailing and Consumer Services, 58*, 102325.

Chang, Y. W., Hsu, P. Y., Chen, J., Shiau, W. L., & Xu, N. (2023). Utilitarian and/or hedonic shopping–consumer motivation to purchase in smart stores. *Industrial Management & Data Systems, 123*(3), 821–842.

Chattaraman, V., Kwon, W. S., Ross, K., Sung, J., Alikhademi, K., Richardson, B., & Gilbert, J. E. (2024). 'Smart'Choice? Evaluating AI-Based mobile decision bots for in-store decision-making. *Journal of Business Research, 183*, 114801.

Chen, C., Zhang, K. Z., Chu, Z., & Lee, M. (2024). Augmented reality in the metaverse market: The role of multimodal sensory interaction. *Internet Research, 34*(1), 9–38.

Crompton, H., & Song, D. (2021). The potential of artificial intelligence in higher education. *Revista virtual Universidad catolica del Norte, 62*.

Dacko, S. G. (2017). Enabling smart retail settings via mobile augmented reality shopping apps. *Technological Forecasting and Social Change, 124*, 243–256.

de Amorim, I. P., Guerreiro, J., Eloy, S., & Loureiro, S. M. C. (2022). How augmented reality media richness influences consumer behaviour. *International Journal of Consumer Studies, 46*(6), 2351–2366.

De Gauquier, L., Brengman, M., & Willems, K. (2020). The rise of service robots in retailing: Literature review on success factors and pitfalls. In *Retail futures: The good, the bad and the ugly of the digital transformation* (pp. 15–35). Emerald Publishing Limited.

Dev, J., & Dev, S. (2023). "How can I help you?": User perceptions of privacy in retail chat agents. In *Extended abstracts of the 2023 CHI conference on human factors in computing systems* (pp. 1–6). Association for Computing Machinery.

DigitalDefynd, T. (2024a, November 28). *5 ways H&M is using AI [case study] [2025] - DigitalDefynd*. DigitalDefynd. https://digitaldefynd.com/IQ/hm-using-ai-case-study/

DigitalDefynd, T. (2024b, December 10). *5 Ways Zara is Using AI [Case Study][2025] - DigitalDefynd*. DigitalDefynd. https://digitaldefynd.com/IQ/ways-zara-using-ai/

DigitalDefynd, T. (2025, March 15). *5 Ways Marks & Spencer is Using AI [Case Studies][2025] - DigitalDefynd*. DigitalDefynd. https://digitaldefynd.com/IQ/ways-marks-spencer-use-ai/

Donnelly, C. (2019, May 15). *Stella McCartney and Google cloud start machine learning and fashion-focused sustainability push*. ComputerWeekly.com. https://www.computerweekly.com/news/252463377/Stella-McCartney-and-Google-Cloud-embark-on-machine-learning-and-fashion-focused-sustainability-push

Đurđević, N., Labus, A., Barać, D., Radenković, M., & Despotović-Zrakić, M. (2022). An approach to assessing shopper acceptance of beacon triggered promotions in smart retail. *Sustainability, 14*(6), 3256.

Ettis, S. A. (2017). Examining the relationships between online store atmospheric color, flow experience and consumer behavior. *Journal of Retailing and Consumer Services, 37*, 43–55.

Fakhry, N. A. M., & Nasr, M. M. (2023). Metaverse and brand experience in fashion digital marketing. *International Design Journal, 13*(3), 331–344.

Farah, M. F., Ramadan, Z. B., & Harb, D. H. (2019). The examination of virtual reality at the intersection of consumer experience, shopping journey and physical retailing. *Journal of Retailing and Consumer Services, 48*, 136–143.

Favaretto, M., De Clercq, E., Schneble, C. O., & Elger, B. S. (2020). What is your definition of big data? Researchers' understanding of the phenomenon of the decade. *PLoS One, 15*(2), e0228987.

Fazal-e-Hasan, S. M., Amrollahi, A., Mortimer, G., Adapa, S., & Balaji, M. S. (2021). A multi-method approach to examining consumer intentions to use smart retail technology. *Computers in Human Behavior, 117*, 106622.

Ferber, T., & Vaziri, D. (2024). Investigating user experience of HRI in the context of a realistic retail scenario-the influence of consumer age on the evaluation of a humanoid service robot. *Journal of Consumer Satisfaction, Dissatisfaction and Complaining Behavior, 37*(1), 187–201.

Fiestas Lopez Guido, J. C., Kim, J. W., Popkowski Leszczyc, P. T., Pontes, N., & Tuzovic, S. (2024). Retail robots as sales assistants: How speciesism moderates the effect of robot intelligence on customer perceptions and behaviour. *Journal of Service Theory and Practice, 34*(1), 127–154.

Fisher, M., & Raman, A. (2018). Using data and big data in retailing. *Production and Operations Management, 27*(9), 1665–1669.

Flavián, C., Akdim, K., & Casaló, L. V. (2023). Effects of voice assistant recommendations on consumer behavior. *Psychology & Marketing, 40*(2), 328–346.

Floh, A., & Madlberger, M. (2013). The role of atmospheric cues in online impulse-buying behavior. *Electronic Commerce Research and Applications, 12*(6), 425–439.

Frank, D. A., & Otterbring, T. (2024). Autonomy, power and the special case of scarcity: Consumer adoption of highly autonomous artificial intelligence. *British Journal of Management, 35*(4), 1700–1723.

Gao, L., & Bai, X. (2014). Online consumer behaviour and its relationship to website atmospheric induced flow: Insights into online travel agencies in China. *Journal of Retailing and Consumer Services, 21*(4), 653–665.

Gümüş, N., & Çark, Ö. (2021). The effect of customers' attitudes towards chatbots on their experience and behavioural intention in Turkey. *Interdisciplinary Description of Complex Systems: INDECS, 19*(3), 420–436.

Guo, W., & Luo, Q. (2023). Investigating the impact of intelligent personal assistants on the purchase intentions of Generation Z consumers: The moderating role of brand credibility. *Journal of Retailing and Consumer Services, 73*, 103353.

Gurusamy, R. (2024). Digital payment technology and consumer behaviour–saving, spending patterns: Are saving and spending patterns a concern? *Journal of Information Technology Teaching Cases, 14*(2), 223–226.

H&M Foundation. (2024, July 21). *AI-powered Smart Garment Sorting System automating textile recycling – H&M Foundation*. https://hmfoundation.com/project/ai-powered-smart-garment-sorting-system-automating-textile-recycling/

Halliday, S. (2022, April 28). Philipp Plein launches crypto concept store, "groundbreaking" NFT platform. *FashionNetwork.com*. https://ww.fashionnetwork.com/news/Philipp-plein-launches-crypto-concept-store-groundbreaking-nft-platform,1400840.html

Han, S. L., An, M., Han, J. J., & Lee, J. (2020). Telepresence, time distortion, and consumer traits of virtual reality shopping. *Journal of Business Research, 118*, 311–320.

Hariharan, A., Pfaff, N., Manz, F., Raab, F., Felic, A., & Kozsir, T. (2020). Enhancing product configuration and sales processes with extended reality. In *Augmented reality and virtual reality: Changing realities in a dynamic world* (pp. 37–50). Springer.

He, X., Liu, Q., & Jung, S. (2024). The impact of recommendation system on user satisfaction: A moderated mediation approach. *Journal of Theoretical and Applied Electronic Commerce Research, 19*(1), 448–466.

Herz, M., & Rahe, V. (2020). Virtual reality becoming part of our lives—Assessing consumers' perceived applicability of virtual reality supported tasks and a critical reflection on the development. In *Augmented reality and virtual reality: Changing realities in a dynamic world* (pp. 113–122). Springer.

Hinmikaiye, A., Adewale, A., & Aladenika, B. (2022). The impact of web 3 on digital marketing strategies of retail businesses in the United States. *Strategic Survey, 10*, 95–04.

Hossain, M. S., Chisty, N. M. A., Hargrove, D. L., & Amin, R. (2021). Role of Internet of Things (IoT) in retail business and enabling smart retailing experiences. *Asian Business Review, 11*(2), 75–80.

Hsieh, J. K., Hsieh, Y. C., Chiu, H. C., & Yang, Y. R. (2014). Customer response to web site atmospherics: Task-relevant cues, situational involvement and PAD. *Journal of Interactive Marketing, 28*(3), 225–236.

Hsieh, S. H., & Lee, C. T. (2021). Hey Alexa: Examining the effect of perceived socialness in usage intentions of AI assistant-enabled smart speaker. *Journal of Research in Interactive Marketing, 15*(2), 267–294.

Huang, N., & Qin, G. (2011, May). A study of online virtual fitting room adoption based on UTAUT. In *2011 international conference on E-business and E-government (ICEE)* (pp. 1–4). IEEE.

İnce, Z., & Kılıç, V. A. (2016). The smartphone usage habits of high school students and their addiction to smart phones. *Uluslararası Eğitim Bilimleri Dergisi/The Journal of International Education Science, 3*(6), 104–123.

Invento RFID LLC. (2024, December 24). Elevating the shopping experience: RFID in fashion stores. *Invento RFID*. https://inventorfid.com/elevating-the-shopping-experience-rfid-technology-in-apparel-stores.

Istiqomah, P., & Alfansi, L. (2024). Navigating style: Exploring the influence of perceived benefit and perceived ease of use on attitude towards use in AI-enhanced fashion E-commerce. *Journal of Entrepreneurship and Business, 5*(1), 1–14.

Ivanov, A., Mou, Y., & Tawira, L. (2023). Avatar personalisation vs. privacy in a virtual try-on app for apparel shopping. *International Journal of Fashion Design, Technology and Education, 16*(1), 100–109.

Jabbar, S., Lloyd, H., Hammoudeh, M., Adebisi, B., & Raza, U. (2021). Blockchain-enabled supply chain: Analysis, challenges, and future directions. *Multimedia Systems, 27*, 787–806.

Jain, S., & Gandhi, A. V. (2021). Impact of artificial intelligence on impulse buying behaviour of Indian shoppers in fashion retail outlets. *International Journal of Innovation Science, 13*(2), 193–204.

Jain, V., & Wah, C. (2022). Computer vision in fashion trend analysis and applications. *Journal of Student Research, 11.*

Jaiswal, N., Misra, A., Khang, A., & Misra, P. K. (2023). The role of internet of things technologies in business and production. In *AI-aided IoT technologies and applications for smart business and production* (pp. 1–13). CRC Press.

Jham, V., Malhotra, G., & Sehgal, N. (2023). Consumer-brand relationships with AI anthropomorphic assistant: Role of product usage barrier, psychological distance and trust. *The International Review of Retail, Distribution and Consumer Research, 33*(2), 117–133.

Jiang, K., Qin, M., & Li, S. (2022). Chatbots in retail: How do they affect the continued use and purchase intentions of Chinese consumers? *Journal of Consumer Behaviour, 21*(4), 756–772.

Jiang, Q., Gu, C., Feng, Y., Wei, W., & Tsai, W. C. (2023). Study on the continuance intention in using virtual shoe-try-on function in mobile online shopping. *Kybernetes, 52*(10), 4551–4575.

Jiang, Q., Sun, J., Yang, C., & Gu, C. (2021). The impact of perceived interactivity and intrinsic value on users' continuance intention in using mobile augmented reality virtual shoe-try-on function. *Systems, 10*(1), 3.

Jin, B., Kim, G., Moore, M., & Rothenberg, L. (2021). Consumer store experience through virtual reality: Its effect on emotional states and perceived store attractiveness. *Fashion and Textiles, 8*, 1–21.

Johnson, N. (2024, April 10). *15+ luxury fashion brands into NFTs in 2023*. NFT Evening. https://nftevening.com/luxury-fashion-brands-nft

Kalyvaki, M., McIntosh, H., & Nash, K. (2023). Virtual selfhood and consumer behavior: Exploring avatar attachment and consumption patterns in Second Life's metaverse. *Computers in Human Behavior: Artificial Humans, 1*(2), 100016.

Kang, H. J., Shin, J. H., & Ponto, K. (2020). How 3D virtual reality stores can shape consumer purchase decisions: The roles of informativeness and playfulness. *Journal of Interactive Marketing, 49*, 70–85.

Kang, J. Y. M., Kim, J. E., Lee, J. Y., & Lin, S. H. (2023). How mobile augmented reality digitally transforms the retail sector: Examining trust in augmented reality apps and online/offline store patronage intention. *Journal of Fashion Marketing and Management: An International Journal, 27*(1), 161–181.

Kannan, M. J., Bhopal, M., & Khan, A. (2024). *Big data analytics unveiled predicting consumer behavior through data-driven strategies for smart retail marketing.*

Kato, B. (2024, October 31). *Fashion models step aside as AI glam bots join them in ads: "It's about faster content creation"*. New York Post. https://nypost.com/2024/10/31/lifestyle/fashion-models-pushed-to-the-side-as-ai-glam-bots-take-over

Kelly, D. (2021, September 21). *Balenciaga partners with Fortnite on In-Game and IRL outfits*. Hypebeast. https://hypebeast.com/2021/9/balenciaga-fortnite-outfits-collaboration-apparel-collection

Khan, A., & Ahmed, A. (2024). Optimizing retail operations, inventory management and sales forecasting with big data and AI in China. *Emerging Trends in Machine Intelligence and Big Data, 16*(1), 18–37.

Kilani, A. A. H. Z., Kakeesh, D. F., Al-Weshah, G. A., & Al-Debei, M. M. (2023). Consumer post-adoption of e-wallet: An extended UTAUT2 perspective with trust. *Journal of Open Innovation: Technology, Market, and Complexity, 9*(3), 100113.

Kim, Y. (2021). Revitalization of offline fashion stores: Exploring strategies to improve the smart retailing experience by applying mobile technology. *Sustainability, 13*(6), 3434.

Kim, W. B., & Choo, H. J. (2023). How virtual reality shopping experience enhances consumer creativity: The mediating role of perceptual curiosity. *Journal of Business Research, 154*, 113378.

Kim, G., Jin, B., & Shin, D. C. (2022). Virtual reality as a promotion tool for small independent stores. *Journal of Retailing and Consumer Services, 64*, 102822.

Kim, H. Y., Lee, J. Y., Mun, J. M., & Johnson, K. K. (2017). Consumer adoption of smart in-store technology: Assessing the predictive value of attitude versus beliefs in the technology acceptance model. *International Journal of Fashion Design, Technology and Education, 10*(1), 26–36.

Kim, H., Choi, Y. J., & Lee, Y. (2015). Web atmospheric qualities in luxury fashion brand web sites. *Journal of Fashion Marketing and Management, 19*(4), 384–401.

Kim, J. H., Kim, M., Park, M., & Yoo, J. (2021). How interactivity and vividness influence consumer virtual reality shopping experience: The mediating role of telepresence. *Journal of Research in Interactive Marketing, 15*(3), 502–525.

Kim, D. E., & LaBat, K. (2013). Consumer experience in using 3D virtual garment simulation technology. *Journal of the Textile Institute, 104*(8), 819–829.

Kim, D. Y., Lee, H. K., & Chung, K. (2023). Avatar-mediated experience in the metaverse: The impact of avatar realism on user-avatar relationship. *Journal of Retailing and Consumer Services, 73*, 103382.

Kim, H., & Lennon, S. J. (2010). E-atmosphere, emotional, cognitive, and behavioral responses. *Journal of Fashion Marketing and Management: An International Journal, 14*(3), 412–428.

Kim, H. Y., Lee, Y., Cho, E., & Jung, Y. J. (2020). Digital atmosphere of fashion retail stores. *Fashion and Textiles, 7*, 1–17.

Kim, N., & Kim, W. H. (2024). The effect of meta-commerce consumption value on consumer satisfaction and continuous usage intention: The moderating roles of meta-verse involvement and consumer competency. *Asia Pacific Journal of Marketing and Logistics*.

Kim, W. B., & Hur, H. J. (2024). What makes people feel empathy for AI chatbots? Assessing the role of competence and warmth. *International Journal of Human–Computer Interaction, 40*(17), 4674–4687.

Koh, C. E., Kim, H. J., & Kim, E. Y. (2006). The impact of RFID in retail industry: Issues and critical success factors. *Journal of Shopping Center Research, 13*(1), 101–117.

Koo, W., & Park, H. (2017). Critical atmospheric cues in designing online stores: The case of Amazon.com. *International Journal of Marketing Studies, 9*(1), 37–45.

Kumar, A., & Shankar, A. (2024). The bold decision to go "all in": Understanding the reasons behind consumers' willingness to migrate to the metaverse. *Psychology & Marketing, 41*(8), 1769–1791.

Lau, K. W., & Lee, P. Y. (2015). The role of stereoscopic 3D virtual reality in fashion advertising and consumer learning. In *Advances in advertising research (Vol. VI) the digital, the classic, the subtle, and the alternative* (pp. 75–83). Springer Fachmedien Wiesbaden.

Le, X. C. (2022). The diffusion of mobile QR-code payment: An empirical evaluation for a pandemic. *Asia-Pacific Journal of Business Administration, 14*(4), 617–636.

Lee, H., & Xu, Y. (2020). Classification of virtual fitting room technologies in the fashion industry: From the perspective of consumer experience. *International Journal of Fashion Design, Technology and Education, 13*(1), 1–10.

Lee, K. Y., Sheehan, L., Lee, K., & Chang, Y. (2021). The continuation and recommendation intention of artificial intelligence-based voice assistant systems (AIVAS): The influence of personal traits. *Internet Research, 31*(5), 1899–1939.

Lee, M., & Park, J. S. (2022). Do parasocial relationships and the quality of communication with AI shopping chatbots determine middle-aged women consumers' continuance usage intentions? *Journal of Consumer Behaviour, 21*(4), 842–854.

Lezama, A. (2025, April 2). Global designers challenge borders, identity, and the limits of physical fashion at Metaverse Fashion Week 2025. *markets.businessinsider.com*. https://markets.businessinsider.com/news/currencies/global-designers-challenge-borders-identity-and-the-limits-of-physical-fashion-at-metaverse-fashion-week-2025-1034542378

Li, Y., & Wang, C. (2022). Effect of customer's perception on service robot acceptance. *International Journal of Consumer Studies, 46*(4), 1241–1261.

Lin, C. Y. (2022). Understanding consumer perceptions and attitudes toward smart retail services. *Journal of Services Marketing, 36*(8), 1015–1030.

Lockwood, L. (2021, December 16). *Tommy Hilfiger Reveals Tommy x Roblox Creators Virtual Collection.* WWD. https://wwd.com/feature/tommy-hilfiger-reveals-tommy-x-roblox-creators-virtual-collection-1235019532/.

Loeb, W. (2022, August 4). *Gucci leads in crypto transactions*. Forbes. https://www.forbes.com/sites/walterloeb/2022/08/04/gucci-leads-in-crypto-transactions/

M&S Sparks. (n.d.). *Scan and shop.* https://www.marksandspencer.com/mysparks/scan-and-shop

Malarvannan, M. (2023, February 15). *Fashion giants go virtual: How lading brands are embracing the metaverse.* Web Analytic. https://analytickit.com/fashion-giants-go-virtual-how-leading-brands-are-embracing-the-metaverse

Manchanda, M., & Deb, M. (2021). On m-commerce adoption and augmented reality: A study on apparel buying using m-commerce in Indian context. *Journal of Internet Commerce, 20*(1), 84–112.

Manis, K. T., & Choi, D. (2019). The virtual reality hardware acceptance model (VR-HAM): Extending and individuating the technology acceptance model (TAM) for virtual reality hardware. *Journal of Business Research, 100*, 503–513.

Marr, B. (2017, September 25). *The amazing ways Burberry is using artificial intelligence and big data to drive success*. Forbes. https://www.forbes.com/sites/bernardmarr/2017/09/25/the-amazing-ways-burberry-is-using-artificial-intelligence-and-big-data-to-drive-success/

Mathews, A. (2024, December 29). *How Nike is using AI to transform product design, customer experience, and operational efficiency*. AIM Research | Artificial Intelligence Market Insights. https://aimresearch.co/market-industry/how-nike-is-using-ai-to-transform-product-design-customer-experience-and-operational-efficiency

Mattan, M. (2024, July 18). *Virtual Try on clothing examples to inspire fashion brands.* https://www.brandxr.io/virtual-try-on-clothing-examples

McDowell, M., & Schulz, M. (2023, April 4). Ralph Lauren's new Web3-centric Miami store accepts crypto. *Vogue Business.* https://www.voguebusiness.com/technology/ralph-laurens-new-web3-centric-miami-store-accepts-crypt

McQueen, S., & Hamza-Lup, F. (2018). *Survey on 3D scanning and Web3D interfaces for the apparel industry* (No. 660). EasyChair.

Mearian, L. (2015, October 8). *Adidas to 3D print custom insoles in sneakers*. Computerworld. https://www.computerworld.com/article/1649469/adidas-to-3d-print-custom-insoles-in-sneakers.html

Medida, R. S. S. (2020). Scope of blockchain technology in the retail industry. *International Journal of Computer Engineering and Technology, 11*(3).

Miao, M., Jalees, T., Zaman, S. I., Khan, S., Hanif, N. U. A., & Javed, M. K. (2022). The influence of e-customer satisfaction, e-trust and perceived value on consumer's repurchase intention in B2C e-commerce segment. *Asia Pacific Journal of Marketing and Logistics, 34*(10), 2184–2206.

Mishra, A., Shukla, A., & Sharma, S. K. (2022). Psychological determinants of users' adoption and word-of-mouth recommendations of smart voice assistants. *International Journal of Information Management, 67*, 102413.

Mohd-Ramly, S., & Omar, N. A. (2017). Exploring the influence of store attributes on customer experience and customer engagement. *International Journal of Retail & Distribution Management, 45*(11), 1138–1158.

Moncler. (2023, October 5). *THE MONCLER x ADIDAS ORIGINALS CAMPAIGN PUTS EXPLORERS IN THE SPOTLIGHT*. Moncler Belgium. https://www.moncler.com/en-be/stories/moncler-x-adidas.html

Mpinganjira, M., Dlodlo, N., & Idemudia, E. C. (2024). Perceived experiential value and continued use intention of e-retail chatbots. *International Journal of Retail & Distribution Management, 52*(13), 121–135.

Muhammad Sohail Jafar, R., Ahmad, W., & Chen, Y. (2024). Metaverse in human behavior: The role of telepresence and flow experience on consumers' shopping behavior in the Metaverse. *SAGE Open, 14*(2), 21582440241261256.

Mythili, J., Menmoli, S. T., Dhanusuya, S., & Sethuram, M. (2024). Intelligent decision-making algorithm for IoT enabled smart shopping. In *2024 international conference on communication, computing and Internet of Things (IC3IoT)* (pp. 1–7). IEEE.

Naqvi, M. H. A., Hongyu, Z., Naqvi, M. H., & Kun, L. (2023). Impact of service agents on customer satisfaction and loyalty: Mediating role of Chatbots. *Journal of Modelling in Management, 19*(2), 470–491.

Ng, S. I., Ho, J. A., Lim, X. J., Chong, K. L., & Latiff, K. (2021). Mirror, mirror on the wall, are we ready for Gen-Z in marketplace? A study of smart retailing technology in Malaysia. *Young Consumers, 22*(1), 68–89.

Ngo, T. K. T., & Nguyen, T. H. (2021). The intention to use QR code payment in an emerging market–the role of "attitude" as mediator. *Psychology and Education Journal, 58*(1), 3440–3454.

Nunkoo, R., Shukla, A., & Mishra, A. (2024). The role of temporal distance, conspicuous consumption, and multisensory interfaces on consumer judgments in metaverse. *Journal of Retailing and Consumer Services, 81*, 104037.

Ogunjimi, A., Rahman, M., Islam, N., & Hasan, R. (2021). Smart mirror fashion technology for the retail chain transformation. *Technological Forecasting and Social Change, 173*, 121118.

Oh, S., Kim, W. B., & Choo, H. J. (2024). The effect of avatar self-integration on consumers' behavioral intention in the metaverse. *International Journal of Human–Computer Interaction, 40*(23), 7840–7853.

Olabiyi, K. (2023, June 24). Fashion data analytics: Unveiling the secrets of fashion's supply chain. *Medium.* https://medium.com/data4fashion/fashion-data-analytics-unveiling-the-secrets-of-fashions-supply-chain-c3da5c8f8acb

Oliveira, E. (2019). Artificial intelligence: An overview. In *Cutting Edge Technologies and Microcomputer Applications for Developing Countries* (pp. 61–65). Routledge.

Oyman, M., Bal, D., & Ozer, S. (2022). Extending the technology acceptance model to explain how perceived augmented reality affects consumers' perceptions. *Computers in Human Behavior, 128*, 107127.

Öztürk, S. (2023, January 1). *Moda tasarımında dijitalleşme | Piyon Co.* Piyon Co. https://www.piyon.co/sayi-1/moda-tasariminda-dijitallesme

Park, A., Wilson, M., Robson, K., Demetis, D., & Kietzmann, J. (2023). Interoperability: Our exciting and terrifying Web3 future. *Business Horizons, 66*(4), 529–541.

Pasquinelli, A. (n.d.). *Metaverse fashion week 2023.* ISTITUTOMARANGONI. https://www.istitutomarangoni.com/en/maze35/community/metaverse-fashion-week-2023

Patil, K., & Pramod, D. (2022, June). Can Metaverse retail lead to purchase intentions among the youth? A stimulus-organism-response theory perspective. In *2022 ASU international conference in emerging technologies for sustainability and intelligent systems (ICETSIS)* (pp. 314–320). IEEE.

Pereira, A. M., Moura, J. A. B., Costa, E. D. B., Vieira, T., Landim, A. R., Bazaki, E., & Wanick, V. (2022). Customer models for artificial intelligence-based decision support in fashion online retail supply chains. *Decision Support Systems, 158*, 113795.

Perumal, S., Jaganathan, M., Shaari, H., & Sulaiman, Y. (2023). Examining perception and attitude towards smart retail technologies in the retail sector. *Russian Law Journal, 11*(7s).

Perumal, S., Qing, Y., & Jaganathan, M. (2022). Factors influencing attitudes and intentions towards smart retail technology. *International Journal of Data and Network Science, 6*(2), 595–602.

Pfeifer, P., Hilken, T., Heller, J., Alimamy, S., & Di Palma, R. (2023). More than meets the eye: In-store retail experiences with augmented reality smart glasses. *Computers in Human Behavior, 146*, 107816.

Pham Thi, T. D., & Duong, N. T. (2025). Consumers' intentions to use AI Chatbots on online shopping platforms. *Journal of Computer Information Systems*, 1–16.

Pillai, R., Sivathanu, B., & Dwivedi, Y. K. (2020). Shopping intention at AI-powered automated retail stores (AIPARS). *Journal of Retailing and Consumer Services, 57*, 102207.

Pillai, R., Sivathanu, B., Rana, N. P., & Struweg, I. (2024a). Assessing the determinants of metaverse adoption for e-commerce retailing. *Journal of Computer Information Systems*, 1–20.

Pillai, R., Sivathanu, B., Rana, N. P., Preet, R., & Mishra, A. (2024b). Factors influencing customers' apparel shopping intention in metaverse. *Journal of Computer Information Systems*, 1–16.

Plotkina, D., Dinsmore, J., & Racat, M. (2022). Improving service brand personality with augmented reality marketing. *Journal of Services Marketing, 36*(6), 781–799.

Prasad, J. P., & Venkatesham, T. (2021). Big data analytics-in retail sector. *International Journal of Computer Science and Mobile Computing, 10*(7), 34–38.

PUMA. (n.d.). *PUMA expands its metaverse web destination, Black Station, with immersive digital retail experience.* https://about.puma.com/en/newsroom/news/puma-expands-its-metaverse-web-destination-black-station-immersive-digital-retail

Rafferty, N. E., & Fajar, A. N. (2022). Integrated QR payment system (QRIS): Cashless payment solution in developing country from merchant perspective. *Asia Pacific Journal of Information Systems, 32*(3), 630–655.

Rahman, T., Hossain, M. A., Kim, Y. S., Noh, M., & Lee, C. K. (2020). Determinants of social media-based online store adoption. In *The 9th international conference on smart media and applications* (pp. 238–243).

Raji, M. A., Olodo, H. B., Oke, T. T., Addy, W. A., Ofodile, O. C., & Oyewole, A. T. (2024). E-commerce and consumer behavior: A review of AI-powered personalization and market trends. *GSC Advanced Research and Reviews, 18*(3), 066–077.

Rehman, U., Shah, M. U., Abbasi, A. Z., & Iqbal, F. (2023). *Social robots in retail: Emotional experiences a critical driver of purchase intention.*

Rese, A., Ganster, L., & Baier, D. (2020). Chatbots in retailers' customer communication: How to measure their acceptance? *Journal of Retailing and Consumer Services, 56*, 102176.

Rhee, H. L., & Lee, K. H. (2021). Enhancing the sneakers shopping experience through virtual fitting using augmented reality. *Sustainability, 13*(11), 6336.

Rohit, K., Shankar, A., Katiyar, G., Mehrotra, A., & Alzeiby, E. A. (2024). Consumer engagement in chatbots and voicebots. A multiple-experiment approach in online retailing context. *Journal of Retailing and Consumer Services, 78*, 103728.

Ronnie, W. (2023, April 27). Zalando launches its innovative virtual dressing room for all its customers. *World of Technology, Video Games & Digital Entertainment.* https://techgameworld.com/zalando-launches-its-innovative-virtual-dressing-room-for-all-its-customers.

Roozen, I., Raedts, M., & Yanycheva, A. (2023). Are retail customers ready for service robot assistants? *International Journal of Social Robotics, 15*(1), 15–25.

Roy, S. K., Balaji, M. S., & Nguyen, B. (2020). Consumer-computer interaction and in-store smart technology (IST) in the retail industry: The role of motivation, opportunity, and ability. *Journal of Marketing Management, 36*(3–4), 299–333.

Roy, S. K., Balaji, M. S., Sadeque, S., Nguyen, B., & Melewar, T. C. (2017). Constituents and consequences of smart customer experience in retailing. *Technological Forecasting and Social Change, 124*, 257–270.

Roy, S. K., Singh, G., Hope, M., Nguyen, B., & Harrigan, P. (2022). The rise of smart consumers: Role of smart servicescape and smart consumer experience co-creation. In *The role of smart technologies in decision making* (pp. 114–147). Routledge.

Ryder, B. (2024, June 11). Which fashion brands used AI best in 2023? *Jing Daily.* https://jingdaily.com/posts/ai-was-the-buzzword-of-2023-which-fashion-brands-used-it-best

Salamone, L. (2023, April 19). How Zegna will use AI to create looks for its customers. *Nss Magazine.* https://www.nssmag.com/en/fashion/32880/zegna-x-ai

Samuely, A. (n.d.). *H&M, Sephora nab early entry into Kik's fashion bot shop | Retail Dive.* https://www.retaildive.com/ex/mobilecommercedaily/hm-sephora-nab-early-entry-into-kiks-fashion-bot-shop

Samuely, A. (2024, October 7). *How Hugo Boss, Prada provide bespoke service with payments*. Retail Dive. https://www.retaildive.com/spons/how-hugo-boss-prada-provide-bespoke-service-with-payments/727454

Saprikis, V., Avlogiaris, G., & Katarachia, A. (2020b). *Investigating the demographic characteristics of university students' perceptions towards their intention to adopt a Mobile augment reality app for entertainment and shopping purposes: A UTAUT model perspective*.

Saprikis, V., Avlogiaris, G., Katarachia, A., & Altini, A. (2020a). Mobile augmented reality apps in shopping malls in Greece: Exploring users' behavioral intentions. *Communications of the IBIMA*, 1–13.

Saricam, C. (2023). Analyzing the influence of store atmospherics on younger generation in apparel retail market with an extended SOR model. *Journal of Global Fashion Marketing, 14*(2), 143–156.

Sarvaiya, M. (2025, February 7). Prada blends Web3 with fashion in its 42nd NFT collection. *The Crypto Times*. https://www.cryptotimes.io/2023/06/01/prada-blends-web3-fashion-in-nft-collection/

Saxena, N., & Sarkar, B. (2023). How does the retailing industry decide the best replenishment strategy by utilizing technological support through blockchain? *Journal of Retailing and Consumer Services, 71*, 103151.

Sen, I. (2018, November 26). *The North Face brings cognitive computing to e-commerce - Outside Insight*. Outside Insight. https://outsideinsight.com/insights/the-north-face-brings-cognitive-computing-to-e-commerce/

Schepers, J., Belanche, D., Casaló, L. V., & Flavián, C. (2022). How smart should a service robot be? *Journal of Service Research, 25*(4), 565–582.

Scholz, J., & Smith, A. N. (2016). Augmented reality: Designing immersive experiences that maximize consumer engagement. *Business Horizons, 59*(2), 149–161.

Schulz, M. (2022, December 20). *2022: A year of fashion NFTs*. Vogue Business. https://www.voguebusiness.com/gallery/2022-a-year-of-fashion-nfts

SDT Inc. (2022, December 19). *11 retailers using IoT for innovative digital transformations*. Medium. https://sdtinc.medium.com/11-retailers-using-iot-for-innovative-digital-transformations-f80e159c473e

Sharma, M., Joshi, S., Luthra, S., & Kumar, A. (2024). Impact of digital assistant attributes on millennials' purchasing intentions: A multi-group analysis using PLS-SEM, artificial neural network and fsQCA. *Information Systems Frontiers, 26*(3), 943–966.

Sharma, S., Islam, N., Singh, G., & Dhir, A. (2022). Why do retail customers adopt artificial intelligence (AI) based autonomous decision-making systems? *IEEE Transactions on Engineering Management, 71*, 1846–1861.

Shaw, N., & Sergueeva, K. (2019). The non-monetary benefits of mobile commerce: Extending UTAUT2 with perceived value. *International journal of information management, 45*, 44–55.

Shehawy, Y. M., Khan, S. M. F. A., Khalufi, N. A. M., & Abdullah, R. S. (2025). Customer adoption of robot: Synergizing customer acceptance of robot-assisted retail technologies. *Journal of Retailing and Consumer Services, 82*, 104062.

Shen, P., Wan, D., & Zhu, A. (2025). Investigating the effect of AI service quality on consumer well-being in retail context. *Service Business, 19*(1), 1–35.

Sidlauskiene, J., Joye, Y., & Auruskeviciene, V. (2023). AI-based chatbots in conversational commerce and their effects on product and price perceptions. *Electronic Markets, 33*(1), 24.

Silva, F. A., Shojaei, A. S., & Barbosa, B. (2023). Chatbot-based services: A study on customers' reuse intention. *Journal of Theoretical and Applied Electronic Commerce Research, 18*(1), 457–474.

Song, C. S., & Kim, Y. K. (2021). Predictors of consumers' willingness to share personal information with fashion sales robots. *Journal of Retailing and Consumer Services, 63*, 102727.

Song, C. S., & Kim, Y. K. (2022). The role of the human-robot interaction in consumers' acceptance of humanoid retail service robots. *Journal of Business Research, 146*, 489–503.

Song, M., Zhu, Y., Xing, X., & Du, J. (2024). The double-edged sword effect of chatbot anthropomorphism on customer acceptance intention: The mediating roles of perceived competence and privacy concerns. *Behaviour & Information Technology, 43*(15), 3593–3615.

Soni, V. (2021). Deep learning and computer vision-based retail analytics for customer interaction and response monitoring. *Eigenpub Review of Science and Technology, 5*(1), 1–15.

Soon, P. S., Gaur, S. S., & Ho, J. S. Y. (2020). Consumers' emotional response to the use of augmented reality (AR): An exploratory study. In *Augmented reality and virtual reality: Changing realities in a dynamic world* (pp. 65–73). Springer.

Staff, A. (2022, June 21). *Look back on 10 years of Amazon robotics.* https://www.aboutamazon.com/news/operations/10-years-of-amazon-robotics-how-robots-help-sort-packages-move-product-and-improve-safety

Statista Research Department. (2025, February 13). *Global spending on blockchain solutions| Statista.* Statista. https://www.statista.com/statistics/800426/worldwide-blockchain-solutions-spending/

Statista. (n.d.). *Internet of Things - Worldwide | Statista market forecast.* https://www.statista.com/outlook/tmo/internet-of-things/worldwide

Statista. (2024a, November 21). *Amount of data created, consumed, and stored 2010–2023, with forecasts to 2028.* https://www.statista.com/statistics/871513/worldwide-data-created/

Statista. (2024b, December 17). *Global virtual fitting room market size 2021–2029.* https://www.statista.com/statistics/1232251/virtual-fitting-room-market-size/

Sung, E. (2020). Brand experience via mobile AR app marketing. In *Augmented reality and virtual reality: Changing realities in a dynamic world* (pp. 3–9). Springer.

Sung, E., Kwon, O., & Sohn, K. (2023). NFT luxury brand marketing in the metaverse: Leveraging blockchain-certified NFTs to drive consumer behavior. *Psychology & Marketing, 40*(11), 2306–2325.

Taggart, E. (2024, November 27). *Robots help John Lewis stack its Christmas gifts high.* The Times. https://www.thetimes.com/business-money/technology/article/robots-help-john-lewis-stack-its-christmas-gifts-high-nl8x7rtlr

Takahashi, D. (2021, December 16). *Forever 21 lets players customize virtual fashion stores in Roblox.* VentureBeat. https://venturebeat.com/games/forever-21-lets-players-operate-virtual-fashion-stores-in-roblox/

Tawira, L., & Ivanov, A. (2023). Leveraging personalization and customization affordances of virtual try-on apps for a new model in apparel m-shopping. *Asia Pacific Journal of Marketing and Logistics, 35*(2), 451–471.

Teles Roxo, M., & Quelhas Brito, P. (2020). "I see myself, therefore I purchase": Factors influencing consumer attitudes towards m-commerce AR apps. In *Augmented reality and virtual reality: Changing realities in a dynamic world* (pp. 51–63). Springer.

Thomas, D. (2021, August 27). *Dolce & Gabbana is bringing NFTs to the traditional world of couture.* British Vogue. https://www.vogue.co.uk/fashion/article/dolce-and-gabbana-nfts-couture

Thormundsson, B. (2025, March 18). *Topic: Artificial intelligence (AI) worldwide.* Statista. https://www.statista.com/topics/3104/artificial-intelligence-ai-worldwide/#topicOverview

Tsai, W. H. S., Liu, Y., & Chuan, C. H. (2021). How chatbots' social presence communication enhances consumer engagement: The mediating role of parasocial interaction and dialogue. *Journal of Research in Interactive Marketing, 15*(3), 460–482.

Tu, M., Wu, L., Wan, H., Ding, Z., Guo, Z., & Chen, J. (2022). The adoption of QR code mobile payment technology during COVID-19: A social learning perspective. *Frontiers in Psychology, 12*, 798199.

Türker, C., Altay, B. C., & Okumuş, A. (2022). Understanding user acceptance of QR code mobile payment systems in Turkey: An extended TAM. *Technological Forecasting and Social Change, 184*, 121968.

Utz, M., Johanning, S., Roth, T., Bruckner, T., & Strüker, J. (2023). From ambivalence to trust: Using blockchain in customer loyalty programs. *International Journal of Information Management, 68*, 102496.

van Berlo, Z. M., van Reijmersdal, E. A., Smit, E. G., & van der Laan, L. N. (2021). Brands in virtual reality games: Affective processes within computer-mediated consumer experiences. *Journal of Business Research, 122*, 458–465.

Violante, M. G., Vezzetti, E., & Piazzolla, P. (2019). How to design a virtual reality experience that impacts the consumer engagement: The case of the virtual supermarket. *International Journal on Interactive Design and Manufacturing (IJIDeM), 13*, 243–262.

Vogue Business in partnership with Stitch Fix. (2024, March 18). *How Stitch Fix is using AI to predict trends*. Vogue Business. https://www.voguebusiness.com/story/events/how-stitch-fix-is-using-ai-to-predict-trends

Vrechopoulos, A., Apostolou, K., & Koutsiouris, V. (2009). Virtual reality retailing on the web: Emerging consumer behavioural patterns. *The International Review of Retail, Distribution and Consumer Research, 19*(5), 469–482.

Ward, G. (2024, November 19). *Could Goddiva's virtual dressing room reduce fashion returns?* Boutique Magazine. https://boutique-magazine.co.uk/could-goddivas-ai-powered-virtual-dressing-room-finally-reduce-returns-for-fashion-retail/

Warden, C. A., Chen, J. F., & Yen, W. H. (2024). Metaverse servicescape: Emotional response to virtual retail design. *International Journal of Human–Computer Interaction*, 1–14.

Weber, F., & Schütte, R. (2019). A domain-oriented analysis of the impact of machine learning—The case of retailing. *Big Data and Cognitive Computing, 3*(1), 11.

Wedel, M., Bigné, E., & Zhang, J. (2020). Virtual and augmented reality: Advancing research in consumer marketing. *International Journal of Research in Marketing, 37*(3), 443–465.

Widayat, W., & Masudin, I. (2023). QR-code-based payment. Does the consumer intend to adopt a retail buying transaction? *Banks and Bank Systems, 18*(3), 1.

Williams, J. (2024, December 19). *The RealReal's luxury resales enjoy boom as demand for new designer items stalls*. Wall Street Journal. https://www.wsj.com/articles/the-realreal-keeps-its-designer-sales-flowing-as-overall-luxury-market-stalls-fe494e2a?

World Fashion Exchange. (2024, December 26). *Top fashion brands leveraging artificial intelligence in 2024*. WFX FASHION & APPAREL BLOG. https://www.worldfashionexchange.com/blog/artificial-intelligence-in-fashion

Xi, N., Chen, J., Gama, F., Korkeila, H., & Hamari, J. (2024). Acceptance of the metaverse: A laboratory experiment on augmented and virtual reality shopping. *Internet Research, 34*(7), 82–117.

Xu, L., Chen, L., Gao, Z., Carranco, L., Fan, X., Shah, N., et al. (2019). Supporting blockchain-based cryptocurrency mobile payment with smart devices. *IEEE Consumer Electronics Magazine, 9*(2), 26–33.

Xue, L., Parker, C. J., & Hart, C. A. (2020). How to engage fashion retail with virtual reality: A consumer perspective. In *Augmented reality and virtual reality: Changing realities in a dynamic world* (pp. 23–35). Springer.

Xue, L., Parker, C. J., & Hart, C. A. (2023). How augmented reality can enhance fashion retail: A UX design perspective. *International Journal of Retail & Distribution Management, 51*(1), 59–80.

Yang, W. (2024). Exploring the impact of humorous chatbots in online retail environments. In *AIP conference proceedings* (Vol. 3220, No. 1). AIP Publishing.

Yuan, C., Wang, S., Liu, Y., & Ma, J. W. (2023). Factors influencing parasocial relationship in the virtual reality shopping environment: The moderating role of celebrity endorser dynamism. *Asia Pacific Journal of Marketing and Logistics, 35*(2), 398–413.

Zara, M., & Hobbs, R. (n.d.). *Decentraland's Metaverse Fashion Week, 2022: Key Takeaways*. https://stylus.com/retail-brand-comms/decentraland-s-metaverse-fashion-week-2022-key-takeaways

Zaytsev, A. (2024, July 8). *Case study: How Levi Strauss & co. is revolutionizing its supply chain with AI*. AIX | AI Expert Network. https://aiexpert.network/levi-strauss-ai/

Zeng, N., Vignali, G., & Ryding, D. (2023). Understanding the application of AI-enabled chatbots in luxury fashion retailing. In *The garment economy: Understanding history, developing business models, and leveraging digital technologies* (pp. 369–388). Springer International Publishing.

Zhang, C., & Lu, Y. (2021). Study on artificial intelligence: The state of the art and future prospects. *Journal of Industrial Information Integration, 23*, 100224.

Zhong, Y., & Moon, H. C. (2022). Investigating customer behavior of using contactless payment in China: A comparative study of facial recognition payment and mobile QR-code payment. *Sustainability, 14*(12), 7150.

Zhong, Z., & Hamouda, M. (2024). The impact of immersive and flow experiences on consumer participation in hyper-connected shopping platforms: A metaverse perspective. *Journal of Consumer Behaviour, 23*(6), 2826–2845.

Zimmermann, R., Mora, D., Cirqueira, D., Helfert, M., Bezbradica, M., Werth, D., et al. (2023). Enhancing brick-and-mortar store shopping experience with an augmented reality shopping assistant application using personalized recommendations and explainable artificial intelligence. *Journal of Research in Interactive Marketing, 17*(2), 273–298.

Index

C. Saricam, N. Okur, *Changing Textile and Apparel Consumption in Transformative Era of Sustainability and Digitalization*, Sustainable Textiles: Production, Processing, Manufacturing & Chemistry,
https://doi.org/10.1007/978-3-031-97419-9

The manufacturer's authorised representative in the EU is Springer Nature Customer Service Centre GmbH, Europaplatz 3, 69115 Heidelberg, Germany. If you have any concerns regarding our products, please contact ProductSafety@springernature.com

Printed and bound by CPI Group (UK) Ltd, Croydon, CR0 4YY
30/09/2026
02225501-0001